Praise for Monte Reel's

Between Man and Beast

"[An] entertaining and provocative story . . . it has the narrative flow and evocative language of a fine historical novel."
—*St. Louis Post-Dispatch*

"[A] sense of urgency compels the reader onward to find out what happened. . . . Arresting."   —*The Washington Post*

"Engrossing . . . would go great with popcorn. . . . Addresses big topics—evolution, abolition—but they remain in service of the narrative, providing context for colorful conflict."
—*The Wall Street Journal*

"A robust intellectual history. . . . In Reel's hands, Du Chaillu's adventures in Africa, including his discovery of Pygmies and his part in a smallpox epidemic, were no less harrowing than his interactions with many of the world's leading scientists and explorers."   —*Publishers Weekly*

"Those unfamiliar with [Paul Du Chaillu] would do well to pick up a copy of *Between Man and Beast*, Monte Reel's new book about Du Chaillu's life and adventures in pursuit of this fierce creature."   —*BookPage*

"Adventure, history, nature, big ideas—what more could you want?"   —*Library Journal*

"Fascinating. . . . A lively footnote to the debate between science and religion and the exploration of the African jungle in the Victorian era."
—*Kirkus Reviews*

"You'd half expect a Bela Lugosi mad scientist or a Johnny Weissmuller Tarzan to pop up in this Victorian-era drama, which travels from the London of Darwin and Dickens to unexplored Africa to Civil War–ravaged America."
—*New York Post*

"A supremely entertaining, enlightening and memorable read."
—*Nature*

"An admirable book for those who like epic tales of exploration. . . . Fascinating."
—*The Buffalo News*

"Retelling his adventures opens a wonderful window, both magical and alarming, into what he [Paul Du Chaillu] saw and, ultimately, into who we are."
—*The Free Lance-Star*

"Swift, clever, well-researched and provocative. . . . Reel skillfully shifts our attention from continent to continent, from past to present, until the story's tributaries merge and rush toward the conclusion."
—*The Plain Dealer*

"A vivid scene worthy of the silver screen. . . . From the perilous adventures to the equally tense academic battles waged by British high society. . . . At times, the mind staggers to recall that this story is a work of nonfiction."
—*San Antonio-Express News*

Monte Reel

Between Man and Beast

Monte Reel is also the author of *The Last of the Tribe.* He has written articles and essays for *The New York Times Magazine, Harper's,* and *Outside,* and he is a former foreign correspondent for *The Washington Post.* He lives outside of Chicago with his wife and two daughters.

www.montereel.com

ALSO BY MONTE REEL

The Last of the Tribe

Between Man and Beast

·=][=·

Between
Man .and. Beast

AN UNLIKELY EXPLORER
and the AFRICAN ADVENTURE
THAT TOOK *the* VICTORIAN
WORLD *by* STORM

· Monte Reel ·

ANCHOR BOOKS

A DIVISION OF RANDOM HOUSE LLC • NEW YORK

FIRST ANCHOR BOOKS EDITION, DECEMBER 2013

The Library of Congress has cataloged the Doubleday edition as follows:
Reel, Monte.
Between man and beast: an unlikely explorer, the evolution debates, and the
African adventure that took the Victorian world by storm/ Monte Reel.—
1st ed.
p. cm.
Includes bibliographical references and index.
1. Du Chaillu, Paul B. (Paul Belloni), 1835–1903.   2. Explorers—Gabon.
3. Hunters—Gabon.   4. Gorilla.   5. Gabon—Description and travel.
I. Title.
DT356.D88R44 2013
916.7210423—dc23
2012014075

Anchor ISBN: 978-0-307-74243-8

www.anchorbooks.com

Printed in the United States of America
10 9 8 7 6 5 4 3 2 1

To Mei-Ling

Contents

PART TWO

PART THREE

A Note to Readers

It was Paul Du Chaillu's luck—good and bad—to have come of age in the late 1850s and early 1860s, when the world was teetering on the sharp edge of transformation. Religious explanations of history, man's place in nature, modern racial conceptions—all were undergoing contentious reassessments that would profoundly shape the coming centuries. Armed with an astonishing collection of zoological specimens and a past full of secrets, Du Chaillu seemed to emerge from nowhere to stumble straight into the center of those debates, helping to push each to unprecedented intensities.

The Victorians might have labeled his story a "Grand Conjunction"—the chance alignment of seemingly disconnected subjects that offered new perspectives on each. That notion has served as a guide for me in writing this book. By fleshing out Du Chaillu's adventures, I'm not attempting to provide a definitive survey of that incredibly fertile moment in history, but I am trying to throw a new angle of light on an era that sometimes feels more familiar than it should.

The book's narrative pivots on the discovery of the gorilla, considered at the time to be man's closest relative and the wildest beast on the planet. I am drawn to the notion of wildness—how it shapes our fears and dreams, and how those fears and dreams can, in turn, reshape the wild. But I never would have stuck with this project if the main attraction had been merely conceptual. The human drama hooked me. It's the story of a nervy young man who rises, and occasionally falls, in a quest to construct a heroic destiny from scratch. That's the heart of this book.

This is nonfiction. Every scene and every quotation is constructed from historical documents. Physical descriptions and atmospheric details are rooted in factual evidence—letters, books, photographs, sketches, memoirs, and newspaper accounts. For those interested in how the narrative was composed, I've tried to be as inclusive as possible in compiling the notes in the back of the book.

One of the satisfactions of writing this book was plunging into the atmospheres of Victorian London at its Dickensian peak, of New York on the verge of the Civil War, and of the African interior at its most lush. It's my hope that the reader might experience a taste of the same pleasure I got when researching the book: the thrill of being swept up by an unknown story and carried away in unexpected directions.

MONTE REEL

Between Man and Beast

··◦)|(◦··

Prologue

He'd been hunting in the forest's depths for months, but he'd never known such silence. No monkeys shook the leaves overhead, no birds cried, no insects droned. The only sounds seemed to come from within: the pulse throbbing in his temples and his own labored breathing.

The previous day the young man had hiked what he guessed was about eighteen miles before collapsing into sleep. But those trails hadn't been nearly as challenging as this one—a muddy ribbon twisting up the forested mountain-side, inset with loose boulders of granite and quartz. He was in good shape and just twenty-five years old, but each step took its toll. He fell behind his companions, whose bare feet gripped the slippery rocks better than did his leather boot soles. His blue cotton shirt and brown pants were streaked with mud.

Somewhere along the way—it was hard to tell exactly where it began—the gentlest of whispers broke through the enveloping hush. The higher he climbed, the louder it got: a breathy hiss that grew into a roar. Twisting through the overgrown vegetation, he found the other men standing on a broad, flat shelf of land. A scene like none he'd ever witnessed burst open in front of him: a vast pool of swirling water, fed by a majestic torrent that spilled down the angled slope for what looked like a mile. A mist rose from the tumult, obscuring everything in a gauzy veil: the swaying ferns, the logs slanting across the water, the trees ringing the banks. According to his calculations, they were about five thousand feet above sea level.

He paused to drink from the pool, but his rest was brief. A short distance uphill, one of his companions spotted footprints that didn't belong to their own party. The feet that had impressed those marks into the mud were bare—but oddly round, with a big toe that seemed to jut away from the other four toes at a severe angle.

When he saw the prints for himself, the hunter felt his heart slam against his rib cage: this was the target he'd traveled so far to pursue, and it finally seemed within his reach.

Following the tracks, the men stumbled into what appeared to be an abandoned tribal village. Years earlier, the land had been cleared for huts that had since collapsed. Stray stalks of sugarcane pushed through the ruins. As the hunter broke off a stalk and sucked the grassy sweetness from its marrow, another of the men observed that some of the plants had recently been ravaged—violently torn up by the roots and mangled into pulp.

They looked at one another and grabbed the rifles they wore strapped across their backs.

More tracks led down a hill. The men carefully crossed a stream on a fallen log, and on the other side of the water they encountered a cluster of enormous granite boulders, some as big as small buildings. The tracks here were even fresher, filled with muddy water that hadn't had time to settle.

The hunter circled to the right of the boulders, while a

few of his companions walked to the left. He emerged from the granite blockade just in time to catch an obstructed view of four dark creatures fleeing rapidly into the dense cover of forest.

The figures disappeared as quickly as they had exploded into view. Running with their heads down and bodies bent forward, the woolly creatures appeared to him, he later noted, "like men running for their lives."

Just minutes before, he might have sworn that the mountain torrent had been the most awe-inspiring sight he'd witnessed in his young life. But this blurred vision of bodies in motion—gone in the blink of an eye—blew it away.

···⊰│ │⊱···

Part One

CHAPTER 1

Destiny

Gabon, West Africa
(Ten years earlier)

Late in 1846, near the end of the rainy season, a group of men reached the Atlantic coast of Africa after weeks of slogging through the waterlogged interior. They had followed no maps, because none existed for that broad swath of equatorial forest. As far as the outside world was concerned, they had emerged from terra incognita—a pure white void in the atlas of the world.

But these men had been exploring the territory all their lives. They were native African traders, and they regularly made long treks from their inland villages toward the largest coastal settlement in Gabon, drawn to the European merchant ships that occasionally dropped anchor to strike deals. On this day, in addition to shouldering the customary bundles of ebony and ivory, the traders carried something extraordinary: a scavenged totem of beguiling rarity.

The American missionary who lived on the bluff wouldn't be able to resist its pull.

His name was John Leighton Wilson, a man of towering stature whose quick smile often got lost inside his fleecy white beard. He had come from South Carolina to the coast of equatorial Africa years before to save the souls of men, but a large part of his own soul had been captured by the wonders that surrounded him. He could spend hours marveling over the elaborate nests of driver ants, or measuring pythons, or trying to tame a porcupine scrabbling near the door of his hut. For all his preaching to the locals about the evils of black magic, false idols, and tribal superstitions, he'd always been vulnerable to the charms of the exotic. And when he spotted the tribesmen's strange curio, he fell under its spell, offering to buy it on the spot.

It was a skull.

At first glance, that calcified mask seemed the product of a peculiarly demented artistry, a grotesquerie of sharp angles and shadowy apertures. When Wilson took it in his hand, it sat heavily, with none of the driftwood airiness of old sun-blanched bones. Its diameter easily exceeded that of a human skull, but there was a passing resemblance, and that's what gave the skull its power to unnerve.

The jaw alone was colossal, framing a mouthful of teeth that seemed to bare themselves in a sinister smile. A quick count revealed thirty-two teeth, the same number as humans, but four of them boldly hijacked Wilson's attention: the twinned sets of upper and lower canines, the largest more than two inches long, curving like scimitars. Those fangs appeared worn, but from *what* he could only imagine.

From the mouth, the facial bones that stretched up toward the eyes sloped back at nearly a forty-five-degree angle, interrupted along the way by a gaping round nasal cavity. From under a menacing ridge of brow, two dark holes stared out where eyes were once socketed. The cranial dome itself was oddly flat, too small in comparison with the rest of the head, as if betraying a brute ignorance that only intensified the promise of danger. The fact that no flesh remained on the skull to provide a more complete picture of the unknown creature's appearance made it no less intimidating: the absence of detail somehow accentuated its eccentricity, in the same way that the most vivid nightmares need darkness to make themselves seen.

The natives called it a *njena*.

Wilson, who for years had been compiling the first-ever dictionary of the local dialect, was unable to translate the term. Whatever the creature was, no words existed for it in English, or in any other language.

WILSON BELIEVED in destiny. Everything and everyone had a place in this world—every grain of sand, the fish of the sea, the fowl of the air, all that creepeth. No matter how pathetic, desolate, abominable, or forlorn something might have seemed at first, its mere existence meant that it was an indispensable part of a divine plan. *And God saw everything that He had made, and, behold, it was very good.*

His faith kept him rooted in West Africa, a place that most of the plantation owners he'd grown up with in South Carolina would have dismissed as an uninhabitable wasteland. In 1842, after a decade in Liberia, Wilson had established the first permanent Christian mission in Gabon. Other churches had staked claims all over Africa, but Gabon—straddling the equator, hanging on to the continent's western edge—remained virtually untouched. European ships had been sailing past the coast for centuries, but the punishing ocean swells and rip currents near the shoreline scared most of them away.

The few that risked disembarking never ventured far inland, because the terrain threw all manner of obstacles in front of would-be explorers. Beyond the narrow coastal plain, the land rose into green hills and then into rugged mountain peaks. The clouds that veiled the highlands dumped more than a hundred inches of rain on the lowland forests. The rivers were choked with mangroves. The mosquitoes were murder. The inland tribes were rumored connoisseurs of human flesh.

Wilson loved it. He built a six-room bamboo house on an airy bluff overlooking the estuary of the Gabon River. The place was called Baraka, a derivative of *barracoon*, the local word for "slave compound," which is exactly what it had been until recently. Wilson hoped to cleanse the ground of its tainted history by doing the work of God. He believed he was fated to live in that very spot.

His wife, Jane, soon joined him. Together they established a Protestant school for the children of the Mpongwe, the ethnic group that populated the loose string of villages on the coast. He thrust himself into their language and their customs. He liked the natives, respecting their "pliancy of character" and what seemed to him a natural quickness of mind. According to Wilson's reading of the Bible, these people were the "descendants of Ham," a son of Noah, and they took a curse with them as they settled in northern Africa after the Great Flood. Whether or not they were cursed, he believed that their continued presence in that harsh landscape proved they were destined to fulfill an important role in history. Wilson tried to get a handle on his thoughts by grabbing a pen and paper:

> *That this people should have been preserved for so long*
> *a period in constantly increasing numbers, and that in the*

*face of the most adverse influences, while other races, who
were placed in circumstances much more favorable for the
perpetuation of their nationality, have passed away from
the earth or dwindled down to a mere handful, is one of the
mysterious providences that admit of no rational explanation,
unless it be that they have been preserved for some important
future destiny.*

Ever on the lookout for providence, Wilson one day in
1848 spotted a small group of Mpongwe tribesmen approaching his hut. They were accompanied by a diminutive figure.
He was a young man, a teenager. He looked like a waif who'd
washed up on a mud bank somewhere upriver, which, according to the story the boy would tell, wasn't too far from the
truth. He had dumped his canoe, he said, and had staggered
along the banks for four days before reaching Baraka. He was
weak, hungry, and utterly pathetic, a shipwrecked soul in
need of safe harbor.

His name was Paul Du Chaillu, and Wilson welcomed
him with open arms. The missionary believed that the boy,
like everyone else, had been born with a divine purpose. It
was Wilson's duty to help him discover it.

No one could have recognized it at the time, but when
the seventeen-year-old crossed the threshold of that bamboo
mission house, he was taking his first step into a new life,
abandoning his old one like the canoe he said he'd let drift
down current. Inside the house, Paul first heard about the
monstrous skull that Wilson had acquired a little more than
a year before. And in that moment, two destinies—the boy's
and the beast's—collided and forever changed course.

————

PAUL SAID he was French, which made perfect sense. Gabon had been claimed as a colony by France just a few years before, a messy process that Wilson and Jane had experienced firsthand.

The captain of a French merchant vessel wandered ashore one day in 1842, a brandy bottle in hand. He summoned King Glass, the local Mpongwe ruler whom the captain had previously met on a trading stop. After the two men drained the jug of brandy, the merchant unfurled a piece of paper and asked the king to sign it.

When the liquor wore off, the king told Wilson that he'd understood the paper was a simple commercial treaty to ensure that trade relations between France and King Glass's monarchy were secure. But the very next day, a French warship fired a salute over Glass's town, and a naval commander triumphantly emerged ashore to inform everyone that they were now officially French subjects. He had a signed agreement to prove it. The commander told them they shouldn't hesitate to call upon the assistance of the French navy if any outsiders, particularly *British* outsiders, attempted to claim their land.

Wilson was away from the house when the commander arrived, but Jane spoke up.

"It is doubtful whether the territory was really ceded," she told the commander, "and the mission does not desire or need French protection."

Her protest, mild but brave, was ignored. The deal had already been done. What's more, it had been replicated up and down the coast. In addition to obtaining the signature of King Glass, the French merchants had already collected signed treaties from his neighboring rulers. It was official: the French navy had been granted legal permission to construct military or commercial outposts wherever it pleased in Gabon.

From the start, France's interest was halfhearted, part of an official "foothold policy" designed to put the brakes on the rapid colonial expansion of England, its longtime rival. Early hopes of turning the coastal plain into a profitable agricultural center quickly fizzled, and construction of a coastal fort was abandoned. The half-completed structure was left to rot in the scouring sea spray as a monument to the country's apathy toward its latest acquisition.

By 1848, the most visible sign of French presence in Gabon was a trading station—the Maison Lamoisse of Le Havre. Paul told Wilson that he had recently arrived in the country from Paris, which was in the throes of a violent revolution, to join his father, who'd moved to Gabon a few years earlier to manage the trading post.

Wilson realized that he in fact had met Paul's father, who had told him about his son a couple of months earlier. The trader, named Charles-Alexis Du Chaillu, had tried to get his son enrolled in a French Catholic mission school nearby, but the Jesuits had refused him entry. The elder Du Chaillu then asked if the Baraka mission might take the boy on, but Wilson couldn't give him an immediate answer. So Paul spent his first months on the coast making short trade runs for his father, traveling upriver to collect goods from tribes that lived several miles inland. He'd been fetching a load of ebony and ivory when his canoe had overturned, he said, forcing him to trek back to Wilson's house.

After their initial meeting, Wilson agreed to take Paul into the school and teach him English, a skill that Charles-Alexis recognized as valuable for an up-and-coming merchant.

Paul leaped at the chance, trading a life that centered on

his father's household for one that revolved around the mission. He became a permanent fixture at the Wilsons' house, moving into one of the boxy rooms that adjoined the "parlor," which the couple had ennobled with an old Windsor chair and some gilt-framed pictures. It was as if Paul had traded his real father for a new one, nabbing a mother as a bonus in the bargain.

Jane Wilson liked the boy as much as her husband did, and she welcomed his cheerful, almost elfin, presence in their home. He exuded energy and optimism and was always quick to dish out compliments—which must have delighted Jane, who proudly clung to a southern belle's sense of style and etiquette. Every day, after she finished tutoring the natives and Paul in their makeshift mission school, she busied herself with the same "delicate little attentions" that she'd observed growing up in Savannah, Georgia: she washed and fixed her hair, putting it up in the style that her husband liked best, and donned a freshly laundered calico dress. For years, she felt as if her husband had been the only person in Gabon to admire her efforts; she never expected the villagers with their square-cut robes to fully appreciate the refinements of a proper Christian lady. But now the unfailingly polite boy was like the doting son she and her husband had never had. Like everyone else, they called him by his first name—a familiarity that Paul would encourage his whole life, among all classes of people, in defiance of the formality of the age.

It wasn't long before he made his feelings about the Wilsons crystal clear: he stopped addressing them as the Reverend and Mrs. Wilson, and he started calling them "Father" and "Mother."

IN THE boy's eyes, Wilson was a miracle: a white man who commanded universal respect among the coastal tribes without resorting to force or coercion, a stately presence who not only *tolerated* his equatorial surroundings but actually regarded them with a kind of devotional reverence. From the start, the boy looked up to Wilson, both literally and figuratively: Paul, standing just over five feet tall, was about a foot shorter than Wilson, whose paternal eminence had earned him a public stature on the Gabon River that rivaled King Glass's for authority.

It wasn't just a reasonable command of conversational English that Paul was picking up from his newly adopted parents. A mix of constant contact and dazzled admiration made him particularly receptive to Wilson's all-embracing immersion in the natural world around him.

Thanks to his trading jaunts, which took him farther inland than others dared, Paul in a few months had already acquired a working knowledge of the region's natural life that far exceeded that of most common traders. But Wilson probably knew more about the flora and fauna of western Africa than anyone alive.

For years, the missionary had been compiling notes for a book that he hoped would chronicle everything worth knowing about western Africa. He'd been the first person ever to study and develop written systems for several of the tribal languages spoken on the coast. He diligently transcribed the local lore, delved into the people's superstitions, and untangled their systems of tribal governance. He mapped the region's rivers and plains, faithfully recorded weather patterns, and attempted to classify nearly every plant and animal he stumbled across, no matter how insignificant it might have appeared. Wilson seemed to affix a literal interpretation to Proverbs 6:6: "Go to the ant, thou sluggard; consider her ways, and be wise." Wilson's entomological observations bore the stamp of an obsessive. He delved into the ants' turreted mounds and dug down into their radiating burrows. He charted the times of day when white ants seemed most active (night), and he timed how fast a swarm of driver ants could consume a live horse or cow (forty-eight hours). He marveled at the way they built arched bridges from one plant to another using nothing but their own bodies and how they collectively formed "rafts" that allowed them to cross streams en masse. These weren't tedious data entries documented with the duty-bound dispassion of a stenographer; they were celebratory appreciations of the diversity of life.

His enthusiasm proved contagious. Paul—whether the tendency was latent in him or planted there by Wilson—blossomed into a keen-eyed chronicler of the inexhaustible wonders that God had created for man. He was particularly fond of the fantastical stories the missionary could tell about the region's more exotic fauna, like the boa constrictor that one day a few years earlier had grabbed one of the Mpongwe's pet dogs. The snake, about halfway through with the process of swallowing the dog whole, had coated the poor mutt's fur from the head down with slimy saliva before Wilson and a couple of tribesmen wrestled it away from the serpent.

"The dog experienced no injury," Wilson said, "but it was several weeks before the varnishing he had got could be removed."

Of all the stories Wilson told, however, none fascinated the boy more than the story of the *njena*. The creature was shrouded in obscurity, spoken of by the locals as if it were a mythical monster, not a real animal. The *njena* was a mystery just waiting to be solved.

·⊰⫘⊱·

CHAPTER 2

A New Obsession

London, England

C aroline Owen tried very hard to maintain a presentable household. The high-backed sofa in the spacious dining room was plush, the leather and stained wood of the library shone with polish, and the colorful carpets in the drawing room left no doubt that the family belonged to Britain's upper crust. But her efforts to preserve decorum often felt doomed. Try as she might, she could not ignore the elephant in the room.

It smelled awful.

"The presence of a portion of the defunct elephant on the premises made me keep all the windows open, especially as the weather is very mild," she confided to her diary. "I got R. to smoke cigars all over the house."

"R." was Richard Owen, the husband responsible for dragging the stinking carcass into their home and one of London's most celebrated men of science. He was always bringing his work home with him, and that work had a way of permeating every room in the house.

If it wasn't an elephant, it was a dead giraffe. Or a hippopotamus. Or a penguin.

Caroline didn't complain too loudly, because she had known from the minute she married Richard exactly what lay in store. As the daughter of a museum keeper, she'd grown up accustomed to the heady aromas of bodily rot. The couple had first met when Richard had been the most promising protégé of Caroline's father. Now, more than a decade later, the couple and their young son lived together in a house on the grounds of the Royal College of Surgeons, where Richard reigned as the most prominent anatomist in England.

The position brought with it a measure of fame. Science as a profession was relatively novel, and it was electrifying the public imagination. The British Empire was sending explorers and naturalists all over the globe, and their biological discoveries almost always ended up on Owen's dissection tables. He liked to coin new terms for the curious items he examined ("dinosaur," for example, was a term he invented), and he wasn't shy about grabbing glory from field scientists; finding a new species was one thing, but classifying that species and placing it into a broader scientific context was something far greater, in his view. Newspapers covered his lectures, and illustrators caricatured his strong chin and his bulging orbs, which scrutinized the world with a bug-eyed intensity.

The public notoriety he'd earned opened the doors to London's highest social circles, and he stormed in with the same unstoppable fervor he'd brought to his science. When a new production of a Weber opera debuted in London and captured his fancy, Richard donned his formal wear to attend the same performance for thirty consecutive nights. He met Charles Dickens backstage in the green room during a performance of *As You Like It*, and the novelist soon became a regular visitor to the Owen house. Queen Victoria and Prince

Albert counted the Owens as friends, and they asked Richard to tutor their children in biology. Lord Tennyson would visit their home to recite poetry in the parlor.

Caroline understood that their lifestyle depended on Richard's work, pungent as it was, and she tolerated her husband's eccentricities with a stiff upper lip. He'd recline on the sofa, smoking one of the cigars he kept in the skull of an Australian aborigine, and she'd take down his dictation. At dinner, he might specially request leg of fowl, just so he could examine the muscular structure of the bird while he dined with her. Before bed, she might tuck in his pet tortoise under a flannel blanket, then pull out her diary and note with wry exasperation that her young son, after watching his dad dissect a chimpanzee in the house, "himself smelt like a specimen preserved in rum."

The last week of April in 1847 unfolded like many others in the Owen house. Richard—battling a stomachache that Caroline blamed on ostrich meat that he'd gamely tasted—was spending much of his time looking at the latest zoological treasure: a squirming, eyeless, flesh-colored amphibian that snaked around blindly in a jar. It looked as if it had been pulled from the lightless depths of a cave, which, in fact, it had. Richard called it a proteus (today it's called an olm), and he persistently tried to feed it worms. Not knowing that such creatures can live up to ten years without food, Owen worried the animal might die of starvation. But his preoccupation with the wriggling creature subsided when a letter was delivered to him that pushed this, and all other preoccupations, out of his mind.

The address of the sender read: "Protestant Mission-House, Gaboon River, West Africa."

————

THE LETTER was signed by Thomas Savage, an American missionary stationed in Liberia who had recently visited John Leighton Wilson in Gabon.

"I have found the existence of an animal of an extraordinary character in this locality," Savage wrote to Owen, "and which I have reason to believe is unknown to the naturalist."

Savage explained that he'd been on his way home to America when his ship was forced to stop in Gabon. While staying with Wilson, he saw the strange skull, and he was captivated. With Wilson's help, Savage begged the native traders for any additional evidence related to the *njena* that they could find. Eventually, the two men collected more skulls and several other bones. The wife of an English missionary who happened to be visiting Wilson at the same time drew anatomical diagrams of the skulls. The sketches were included in the letter that Owen now read.

Savage wrote to request that Owen compare the descriptions of the skulls with those of other species stored in the museum at the Royal College of Surgeons. He also mentioned to Owen that he hoped to procure the carcass of a *njena* and preserve it in alcohol, but was careful not to excite Owen with a promise. "Great uncertainty however attends my success, as they are indescribably fierce and dangerous and are found only in the far interior," Savage added.

Owen was intimately familiar with all the known species of chimpanzees and orangutans, having dissected each multiple times. But as he eyed the drawings, the differences between this skull and those of the other apes popped out in sharp relief. The facial angle wasn't the same as in the chimpanzee, and the nasal bones were more prominent than those of the orangutan. The *njena* appeared to be a different animal altogether.

A naturalist in Bristol whom Owen knew well contacted the captain of a trading vessel, and he asked him if he might try to collect more skulls during an upcoming visit to Gabon. The captain returned a few months later with three skulls, but the man died almost immediately upon his arrival in Bristol, unable to offer any more information about the animal. It didn't matter; Owen had gotten his hands on a real skull, and he wasted no time in giving it a thorough examination.

By February 1848, Owen was ready to deliver a paper to the Zoological Society titled "On a New Species of Chimpanzee." He cataloged all the details he was able to discern—the skull's facial angle, its dental structure, and other osteological traits. As a tribute to the man who'd brought the animal to his attention, Owen proposed to name the new species *Troglodytes savagei*.

But Savage, who'd since returned to the United States, had delivered his skulls and bones to Dr. Jeffries Wyman, an anatomist at Harvard Medical School. Because it generally took months for scientific news to travel across the Atlantic, Owen wasn't aware that Savage and Wyman had already jointly published the first-ever description of the species in the *Boston Journal of Natural History*. The name they gave it was *Troglodytes gorilla*.

Owen had come in second place in the race to name the new species. Conceding the victory, he withdrew the name he'd suggested. But he couldn't stop thinking about this new animal. When something captured Owen's imagination, whether it was an opera or the discovery of an unknown species, it was very difficult for him to let it go.

CHAPTER 3

Hanno's Wake

Gabon

The name "gorilla" was plucked from an ancient Greek text called *The Voyage of Hanno,* written in the fifth century B.C. The document charts the sea journey of Hanno the Navigator, who was ordered by Carthage to lead a fleet of ships beyond the Pillars of Hercules to explore an area that would later be known as northern Africa.

Hanno wrote that he sailed past coastal rivers that teemed with crocodiles and encountered an island that burned with fires at night and was filled with the "sound of pipes, cymbals, drums, and confused shouts." Frightened away from the island, he continued to sail south. He wrote:

> *On the third day after our departure thence, having sailed by those streams of fire, we arrived at a bay called the Southern Horn; at the bottom of which lay an island like the former, having a lake, and in this lake another island, full of savage people, the greater part of whom were women, whose bodies were hairy, and whom our interpreters called* Gorillae. *Though we pursued the men, we could not seize*

any of them; but all fled from us, escaping over the precipices,
and defending themselves with stones. Three women were
however taken; but they attacked their conductors with their
teeth and hands, and could not be prevailed on to accompany
us. Having killed them, we flayed them, and brought their
skins with us to Carthage. We did not sail farther on, our
provisions failing us.

What sorts of beings Hanno actually saw remains any-one's guess, but it almost certainly wasn't the same species of creature whose skull Wilson had acquired. But perhaps the *njena* could have been the same mysterious manlike ani-mal that Andrew Battell—an English sailor imprisoned by the Portuguese in Angola for eighteen years—had described as the "pongo" in the early seventeenth century. Battell, who never saw one himself, described some of the fantastical local legends about the animal:

They goe many together and kill many negroe that travaile
in the woods. Many times they fall upon elephants which come
to feed where they be, and so beat them with their clubbed fists
and pieces of wood that they will runne roaring away from
them. The pongos are never taken alive, because they are so
strong ten men can not hold one of them.

Another traveler, T. E. Bowditch, wrote a book of his travels along Africa's coast in 1819 and described a beast the Mpongwe called the *ingena*. He was told it stood about five feet tall with broad shoulders that spanned four feet. "Its paw was said to be even more disproportioned than its breadth," Bowditch reported, "and one blow of it to be fatal." He also said that numerous natives had told him—"without

variation"—that this anthropoid ape built primitive imita-
tions of their bamboo houses and slept outside on top of the
roofs.

Savage and Wyman tracked down all these hazy historical
references, while Wilson continued to help them collect more
anecdotal information in Gabon. As Du Chaillu hovered close
by, Wilson labored on his encyclopedic book about western
Africa, pressing the Mpongwe natives for whatever they could
tell him about an animal that almost no one near the coast, it
seemed, had ever seen with his own eyes.

Word quickly spread deep into the interior that the men
at the mission house had a keen interest in anything related
to the *njena*, and more skulls and bones began arriving at the
house on the bluff. Eventually, Wilson collected two complete
gorilla skeletons, shipping one to Wyman in the United States,
the other to Owen in London. Wilson even got a glimpse of a
recently slain gorilla, although the corpse had already deterio-
rated to an unsalvageable condition. Between his own observation and the descriptions of the beast provided by a handful
of hunters who'd claimed to have seen one alive, Wilson could
roughly sketch the outline of a terrifying monster.

"It is almost impossible to give a correct idea, either of
the hideousness of its looks, or the amazing muscular power
which it possesses," Wilson wrote.

The natives told him that if the animal met a single person in the forest, it would invariably attack. Wilson said that
he'd seen a man who chanced upon one of the brutes and
barely lived to tell about it. The man's lower leg was mangled,
Wilson said, and it likely would have been torn clean off if his
hunting companions hadn't come to his rescue, just in time.
"It is said they will wrest a musket from the hands of a man
and crush the barrel between their jaws," Wilson wrote, "and

there is nothing, judging from the muscles of the jaws, or the size of their teeth, that renders such a thing impossible."

Using information that Wilson helped them collect, Savage and Wyman wrote that local tribes considered the slaying of a gorilla to be the ultimate act of courage and skill. The Mpongwe, who sometimes enslaved the members of other tribes that lived deeper in the interior, had recounted to Wilson the story of one lowly servant who had shot and killed two gorillas with a rifle. "This act, unheard of before, was considered almost superhuman," Savage and Wyman wrote. "The man's freedom was immediately granted to him, and his name proclaimed abroad as the prince of hunters."

For reasons he kept to himself, Paul was also looking for liberation, a way to permanently break free from a past that he never discussed. But as the stories he was hearing from the natives proved, even the lowliest, most unassuming of men could become legends, worthy of respect. All that was needed was courage and the opportunity to prove it.

Did Paul have to accept the perfunctory fate of becoming a coastal merchant like his father before him? Was he doomed to see the world through the limiting lens of simple commerce, or could he view it in all of its kaleidoscopic splendor? What was stopping him from engaging life on a grand scale?

Wilson fed Paul's dreams of self-transformation. As an African missionary, he was well versed in the story of Moses in the land of the Pharaoh: a baby rescued among the bulrushes of the river Nile, spared the fate of a cursed birth, destined to grow into the prototype of the epic hero. Paul acted as if his own life began when he was rescued from the river and met Wilson. Whatever came before didn't matter. He'd been given the chance to write his own future, and he was eager to start immediately.

Wilson backed him fully, because that was what he believed mission work was all about: giving people the opportunity to follow paths that hadn't been open to their forebears, in the hopes that there they would find their true purpose.

Early in 1852, Wilson discovered that a seminary in Carmel, New York, was looking for someone who could teach French. The wide-open promise of America seemed perfect for a young man determined to erase all antecedents. Wilson sent some letters, pulled some strings. Paul soon met a captain who offered him free passage across the Atlantic. He was off, embarking on what he hoped would be the first chapter of an epic fable of his own making.

·◦][◦·

CHAPTER 4

Drawing Lines

London

I t was impossible to know what morning would bring to the Owen house, but it was safe to guess it might come early. A few of Richard's scientific friends possessed a passion for discovery that equaled his own, blinding them to all manner of social norms, such as observing the clock.

"Mr. Darwin was here very early," Caroline noted in her diary one day in 1847, "before breakfast."

The thirty-eight-year-old Charles Darwin was a regular visitor to Owen's house in 1847 and 1848. The two men had known each other for years; Darwin had sent Owen the mammalian fossils he'd collected during his South American voyage on the HMS *Beagle* between 1831 and 1836. Now Darwin was picking the brain of the senior scientist, trying to wrap his head around an idea that Owen was developing concerning a "unity of organization" common to all vertebrates, including human beings.

Owen believed different species shared a common, fundamental blueprint—an "archetype," as he called it—that had been preprogrammed to develop in a distinct way. In every species, he wrote, "ends are obtained and the interest of the animal promoted, in a way that indicates superior design, intelligence and foresight; but a design, intelligence and fore-sight in which the judgement and reflection of the animal never were concerned; and which, therefore, with Virgil, and with other studious observers of nature, we must ascribe to the Sovereign of the universe, in whom we live, and move, and have our being."

Owen's writings on the subject were both erudite and opaque, which is why Darwin kept popping by his house, trying to more clearly understand his work. As Owen was pre-paring a journal article outlining the concept in 1848, Dar-win pleaded with him to include illustrations so that the ideas might be more intelligible to "ignoramuses such as myself."

Darwin's eager interest and humility flattered Owen. He didn't know that the younger scientist was working on his own theory of biological development that was far more coherent and fastidiously tested than his own. That idea—and Owen's resistance to it—would eventually tear their friend-ship to shreds.

As a member of the Anglican Church, Owen would have kept in his library a Bible that included a time line printed in the margins. The day of creation, it stated, occurred on October 22 in 4004 B.C.

The chronology had been devised by a seventeenth-century Anglican archbishop named James Ussher, and it was based on readings of the Bible, Talmud, and other sacred doc-uments in the Judeo-Christian tradition. Owen dismissed the time line, which for nearly two hundred years was included in the Authorized Version of the King James Bible used by the

Anglican Church. To him and nearly every other scientist of his day, the Bible wasn't a book to be read literally but rather an allegorical or metaphorical text.

The scientific evidence for questioning the Bible's time line became unavoidably apparent to scientists in the late eighteenth and early nineteenth centuries, thanks mainly to discoveries in geology and paleontology. The earth appeared to be far older than the biblical accounts suggested, and it had endured multiple geological epochs. The French naturalist Georges Cuvier and the British geologist William Smith had analyzed fossil remains and geological strata to show that each of those epochs featured distinct species of plants and animals, many of which seemed to no longer exist anywhere in the world. Evidence of human life, as Cuvier noted, appeared only within the most recent geological deposits. The idea of a single moment of creation, during which all species roamed the earth concurrently, began to lose credibility among naturalists. By the first two decades of the nineteenth century, most scientists agreed that new species both appeared and disappeared over time.

Varying explanations sprang up among those who wanted to restore authority to the biblical version of creation. In fact, many of the scientists behind the discoveries that challenged the Scriptures were the same ones who worked hardest to reconcile the findings with biblical accounts. Some proposed a "gap theory," which said an undocumented amount of time existed between "the beginning" of Genesis and the creation of all species; others suggested that each of the six days of creation, as described in the Bible, was much longer than the twenty-four-hour days of our present age. These scientists weren't ready to undermine a religion that they themselves followed faithfully.

For a nineteenth-century man of science like Cuvier, who couldn't ignore the obvious discrepancies, keeping the faith required a reconsideration of what was truly sacred and what wasn't. The biblical time line was open to interpretation. But despite increasing fossil evidence supporting "transmutation," as evolution was most commonly termed in those days, Cuvier rejected the notion. Evolution was one line that couldn't be crossed. "All species reproduce according to their kind," Genesis stated. For Cuvier, this was an untouchable truth. To suggest otherwise seemed both illogical and immoral.

The idea that the defining characteristics of a species, particularly the human species, might change radically over time is what really riled literal-minded Christians, then and now. Unlike many other religious traditions, the Judeo-Christian texts insist upon a single moment of creation. More important, they draw a clear, bold line separating humans from other animals. Man was a special creation, fully formed from the start, wholly separate from and superior to all the other creatures on earth. Humans are absolutely unique, according to the Bible, and always have been.

Most early theories of transmutation tried to preserve the notion of man as the ultimate creation of a divine watchmaker. *Vestiges of the Natural History of Creation*, a book authored by Robert Chambers but published anonymously in 1844, outlined the idea that animals develop from lesser to "more perfect" beings, with humans being the most perfect of all. By following this progressive law of development, Chambers wrote, transmutation was part of a master plan laid out by the "Almighty Author."

The book, regardless of its attempt to appease religious leaders, was condemned as heresy by the conservative members of England's theological and political elite. In need of someone to poke holes in the book's science, members of the Victorian establishment turned to a man they knew they could trust to support the status quo: Richard Owen.

Roderick Impey Murchison, the president of the Royal Geographical Society and a man who would be knighted by Queen Victoria within a year, was one of Owen's closest friends. He wrote to Owen in 1845 suggesting that he attack *Vestiges* because "a real *man in armour* is required, and if you would undertake the concern you would do infinite service to *true* science and sincerely oblige your friends."

But Owen declined. He saw much to admire in the book, and he had slowly come around to accepting the idea of transmutation of species—with limits. His ideas about archetypes were, in a sense, built on the same principles that Chambers outlined. Owen resisted the idea that evolution could result from outside forces, such as Darwin's idea of natural selection. Like Chambers, Owen believed the changes were internally triggered, from within the living beings themselves.

Owen's refusal to attack the *Vestiges* didn't mean that he was turning his back on the establishment. Instead, he seemed to be trying to nudge it toward a position of compromise that accepted new discoveries without wholly abandoning Scripture. Over the next decade, Owen worked to draw a precise line defining just how far he believed the concept of transmutation could be taken: nature, no matter how slow the processes, could never gradually turn an ape into a man.

CHAPTER 5

American Dreams

Carmel, New York

Carmel, nestled between rolling hills, sat a few miles shy of the halfway mark along the telegraph line that connected New York City to Albany. A large pond, called Shaw's Lake, served as a glassy sort of village green for the hamlet's twenty-three hundred souls, and near one of its banks a narrow bridge stretched over the recently laid tracks of the New York & Harlem Railroad. By following the bridge road, a visitor ran straight into the town's most imposing edifice: a hulking mass of pale stone and Doric columns called the Carmel School.

Built just two years before Paul arrived in the United States, the school had floundered as a girls' seminary, with an enrollment of only a dozen or so students. But a new director was determined to turn the school into a thriving institution worthy of its venerable premises. In 1852 he opened the school to both boys and girls and augmented the staff to accompany the swelling ranks. One of his priorities had been to hire an authentic Frenchman as a foreign-language instructor.

Paul seemed to fit the bill. Though he'd learned a passable amount of English from the Wilsons, he spoke it with a soupy Parisian accent, and often, when struggling to find *le mot juste*, he filled the blanks with French. A more self-conscious speaker might have chosen silence over risking the appearance of eccentricity, but Paul just let it fly. He was a verbal caricature, the kind who could be crudely imitated by turning "the" into *dzee* and "these" into *dzees*. In small-town America, he couldn't have drawn more attention to himself if he had tried. That's why few picked up on this sliver of mildly tragic irony: the new faculty member everyone called "the Little Frenchman" wanted nothing more than to be considered a real American and to blend in with everyone else in red-white-and-blue Carmel.

He told his students that he despised France. He said that he'd lived through the revolution of 1848—an experience that forever tainted his view of the Second Empire. He told them he was determined to become a naturalized U.S. citizen, and in fact he visited the Putnam County Courthouse in Carmel to apply. But he didn't wait for the papers to come through to adopt the country as his own. He changed the pronunciation of his last name—which in French sounded kind of like *du-sha-yu*—and happily incorporated a common stateside mispronunciation, turning it into *du-chally-yu*. As far as he was concerned, he was American through and through.

EVEN IF his students believed he was a citizen, it didn't mean they accepted him as one of their own. The same transparent sense of wonder that Wilson had encouraged in Paul left him wide open to ridicule among students and other faculty members who were predisposed to belittling the quirky.

One of his favorite students, Helen Evertson Smith, later

described Paul's days at Carmel School as those of an unwitting innocent who cheerfully reached out for friends by sticking his hand into a viper's nest: "Mr. Du Chaillu's diminutive size, his often exceedingly queer English, his very acts of kindness were all openly ridiculed, not only by boys and girls big enough to know better, but also by the other teachers."

He had falsely added five years to his age, claiming to be in his mid-twenties when he arrived at Carmel, but that didn't seem to fool the students, who afforded him none of the respect of an elder. Once, a group of particularly tyrannical male students devised a detailed plan to kidnap Paul while he slept in the dormitory. They decided to sneak into his room and tie him up in a blanket. Then they would toss him hand to hand, like construction workers moving sandbags, down four flights of stairs, across a snowy lawn, down a steep hillside to Seminary Hill Road, across the bridge, and into the frigid waters of Shaw's Lake. The boys marked a date on the calendar: a suitably chilly one in February. But the day prior to Paul's planned kidnapping, the French teacher arrived at the door of the school with a wagon brimming with treats—cakes, candies, pies, fruits, even turkeys. That day—for no apparent occasion—Paul's classroom on the top floor of the school became the site of an impromptu celebratory feast. Stricken with guilt, the boys abandoned their plan, which was later revealed by one of the youngest of the conspirators. Some suspected that Paul had been warned of the plot and cunningly staged the feast to curry favor and humble the guilty. Others believed he hadn't known a thing, and they argued that the party was completely in character with his generous, exuberant personality. No one ever really knew for sure.

What was certain, however, is that if Paul won the respect of some of his students, his stories of his time in Africa helped him do so. Though he might have desired to be considered a normal American, he learned to use his aura of the exotic to his advantage. The more he told students about his time in Gabon, the more he witnessed the enthusiasm it inspired in some of them. He'd speak of elephants, of hippopotamuses, of boa constrictors—the wilder the beast, it seemed, the greater the interest. A small group of girls, which included Evertson Smith, used to linger in his room after class to spend part of their lunch hour listening to his adventures. He turned it into a sort of exercise, one that the girls grew to cherish: if they asked him questions about Africa in French, he'd respond in English. The exchanges on occasion veered into the personal, as the girls tried to get a clearer picture of the young teacher's backstory. The details he gave them were spare but satisfied their curiosity: he told them he'd been born in Africa and that his mother had died when he was too young to remember her. He was sent to France to attend school, he said, and when he was old enough to work for his father, he rejoined him in Africa.

While at Carmel, Paul capitalized on the interest he noticed that some people harbored for Africa, and he offered to write a series of articles about the region for the *New-York Tribune*. The editors agreed to publish several pieces, which described the coast's animal life.

One of the *Tribune*'s readers, John Cassin, took note of the articles and tracked down their author. Cassin was head of Philadelphia's Academy of Natural Sciences and one of America's most accomplished ornithologists—later he'd be considered by many to be the country's first true taxonomist—and he'd been compiling lists of species of birds found all over the world, from the American West to Japan to Chile. Cassin's interest exhilarated Paul, who was thrilled to have attracted the attention of a learned body as august as the academy. The idea of a return trip to Africa—but this time with the added status that American connections gave him—seized the young man. His mind reeled at the potential zoological treasures a true expedition into Gabon might yield. The number of bird species he'd be able to shoot, stuff, and ship back to Philadelphia would dramatically expand Cassin's inventory of African birds. When he aired the idea to Cassin, the ornithologist heartily encouraged him. If Paul launched such an expedition, Cassin said, the academy would sponsor him.

Thus, on October 16, 1855, Cassin stood before the members of the academy and announced that Paul was about to return to western Africa, undertaking an expedition that would provide them with specimens. During that meeting, the group appointed a committee to solicit contributions to help finance a trip of undetermined length.

It instantly became clear to Paul that his stay in America would not be the wholly transformative experience he at first hoped it might be, but rather a meaningful parenthesis between two chapters in Gabon. In October, he boarded a three-masted schooner in New York and sailed for Africa.

Neither Cassin nor anyone else seemed to know that Paul had his sights set on something far greater than mere birds. He told himself that he wasn't coming back until he had bagged the beast that until that moment had existed more in myth than in reality, a demon-like creature whose killing could earn anyone, even the lowliest slave, the sort of esteem powerful enough to break the chains of any circumstance. Paul wanted a gorilla.

--∘] [∘--

CHAPTER 6

To Slide
into Brutish Immorality

London

R ichard Owen grabbed six different artifacts from the museum of the Royal College of Surgeons. Three were human skulls: an aboriginal from Van Diemen's Land, a Mongolian from central Asia, and a Caucasian from Europe. The other three were the skulls of apes: an orangutan, a chimpanzee, and a gorilla. Throughout the winter of 1854 and 1855, he pored over every detail of each skull, determined to chart their similarities and differences.

Chimpanzees had been known to European scientists since 1699, when an English anatomist dissected a specimen collected in West Africa, and orangutans had first been described by an English traveler to Borneo in 1712. A few decades later, when Carl Linnaeus developed the biological classification system that is still used today, he labeled humans as primates, along with the apes. He believed humans to be essentially unique, and he knew grouping men with apes would trouble people. "It is not pleasing to me that I must place humans among the primates," he wrote to a friend in

1747, "but man is intimately familiar with himself. Let's not quibble over words. It will be the same to me whatever name is applied. But I desperately seek from you and the whole world a general difference between men and simians from the principles of Natural History. I certainly know of none. If only someone might tell me one!"

Within thirty years, scientists came up with something. They separated the primates into two genera—one for four-handed species (apes and monkeys), and another for two-handed ones (humans). By the early nineteenth century, Georges Cuvier had separated them even further, creating a distinct order for humans.

Many scientists today believe that apes shared a common ancestor with humans several million years ago before splitting off on separate evolutionary paths. Modern genetic evidence suggests that the chimpanzee is man's closest living relative, and gorillas are a close second. But when Owen compared features of the skulls—the shape of the orbits, the projection of the nasal bone, the grinding surface of the teeth—he concluded that the gorilla was the most anthropoid, or human-like, of all animals. When he presented these findings at a meeting of the Royal Institution in London in February 1855, the gorilla instantly became the standard benchmark for anyone wanting to compare humans with other animals.

At that same meeting, Owen seemed to feel a threat to the fixed taxonomical lines that had been drawn between humans and apes. His contemporaries had created those distinctions on paper, but the increasing chatter about transmutation suggested that many modern scientists harbored doubts about the stability of species. Owen, addressing the crowd behind a lectern at the Royal Institution, attacked the notion by lauding the wisdom of the seventeenth-century Platonist philosophers like Henry More and Ralph Cudworth, who sought to prove that reason and religion were perfectly compatible.

"The present age may be more knowing, but can it truly flatter itself as being wiser, more logical, and less credulous than that of Cudworth and More?" Owen asked his audience.

He told them that More had anticipated that a future generation might want to blur the lines between man and animal, and as far as Owen was concerned, the thoughts of the wise old sage on the matter still rang true. Owen recited a passage from More: "And of a truth, vile epicurism and sensuality will make the soul of man so degenerate and blind, that he will not only be content to slide into brutish immorality, but please himself in this very opinion that he is a real brute already, an ape, satyre or baboon."

In other words, the resemblance between a man and a gorilla was cause for concern only among those uncivilized enough to doubt mankind's essential separation from mere animals.

Owen seemed untroubled by one obvious fact that he failed to mention during his lecture: no one that he would consider "civilized" had ever actually seen a gorilla.

CHAPTER 7

An Awkward Homecoming

Gabon

The captain's charts showed a mazy obstacle course of shallows and shoals, and an occasional westerly sea breeze crisscrossed the water's currents to whip up peril. Newcomers often tensed when they reached this broad estuary of the Gabon River, but Paul knew that the surf was even worse farther down the coast. Given all he and the rest of the men had endured for weeks aboard that cramped schooner, a mild bout of pitching and rolling was a more-than-equal trade for the promise of dry land.

Paul had spent those weeks with a seven-man crew. The four sailors and the cook had bedded down in the stuffy forecastle, but Paul's status as the schooner's only paying passenger earned him preferential treatment: a bunk in the aft cabin alongside the captain and the mate. By the schooner's standards, this qualified as a luxury.

This cruise was supposed to be the relaxing part of his expedition, but little about it was comfortable. Even for someone as small as Paul, it was impossible to stand upright in that cabin, unless he happened to be directly under the tiny skylight. The only real furnishings consisted of the bunks, a small table in the center of the floor, and a wall-mounted cupboard that held the rattling plates and cutlery. When they ate, the three men sat on the same chests that stored their personal belongings.

Paul had spent most days up on the deck, where the wind whistled through the rigging. If he looked hard enough, something could usually be found to break the monotony of the waves. It might have been a swordfish knifing through the water, or a teeming school of porpoises, or a desperate bullfinch that appeared as if it might drop dead from exhaustion at any moment, so far from the shore. It had been easy to sympathize with the bird: all transatlantic voyages that relied on only currents and wind were endurance tests. The Doldrums, a dead zone of low atmospheric pressure near the equator, stalled their progress for a full week.

"For five days," Paul wrote, "two empty flour barrels that had been thrown overboard remained alongside of our ship."

Eventually, an afternoon squall pushed them into the southeast trade winds. More storms converged on the small ship. Water lashed the bulwarks and threatened to bury the prow, tossing the boat so fiercely that the sailors had to tie themselves to the masts.

In that storm, Paul had thought about the cargo he'd loaded into the hold back in New York, which reminded him of the reason he'd endured the journey in the first place: to explore and to hunt. Every crash of thunder seemed to pose a troubling question: What if one of the glowing branches of lightning somehow connected with the expedition supplies that Paul had stowed in the hold?

"I had enough powder on board to blow the ship to pieces," he later wrote.

———

SHORTLY BEFORE Paul arrived on the coast of Gabon in late 1855, the Wilsons returned to America, a permanent move prompted by a worsening liver ailment John had developed after more than twenty years in West Africa. One of the Mpongwe natives told Paul how they all had gathered around the boat to watch the couple leave the continent. As the man recalled that parting scene, tears welled in his eyes. The locals had grown fond of Wilson, and they missed him.

But the old man's presence thoroughly pervaded Baraka, where Paul planned to stay with Wilson's successors for five months, until the dry season commenced and the swamped interior might be traversable. The saplings Wilson had planted on the bluff had grown into a shady orchard, rich with the scent of tropical fruits. The cadence of life was reassuringly predictable: the stirrings of the fowl house in the early morning; the breakfast pots clanging in the kitchen hut; the chime of the morning school bell summoning the children from the village below; young voices reciting prayers in the Mpongwe language. Paul could amass supplies and plan his route here, comfortable within an atmosphere that provided a constant reminder of his old friend and mentor.

Wilson's departure wasn't the only significant event that immediately preceded Paul's arrival. Around the same time, his father died. Charles-Alexis Du Chaillu was interred near the estuary in a burial ground cluttered with trees and crooked crosses. Whatever feelings the news stirred within his son were lost to history, because Paul never committed a single word to paper on the subject and none of his acquaintances ever publicly recalled him speaking of it.

The only mentions of his father in Paul's writings that

survive from this time are connected with a desire to move *beyond* the memory and influence of his father. Many of the locals assumed that he'd returned to take his father's place as a merchant, and they greeted him warmly, eager to see what he'd brought to trade.

"Their disappointment was great, therefore, when I was obliged to inform them that I had come with no goods to sell," he wrote, "but with the purpose to explore the country back, of which I had heard so many wonderful stories from them, and to hunt wild birds and beasts. At first they believed I was joking."

Some suspected that his exploratory hunting trip was a pretense, a cleverly disguised attempt to circumvent Mpongwe control of the coastal trade. Over centuries of regular visits from Portuguese and Dutch ships, members of the tribe had established a rigid commercial system designed for their benefit. The Mpongwe on the coast, led by King Glass, would procure many of their goods—ebony, ivory, barwood, and even slaves—from other tribes that lived farther inland. If, for example, King Glass traded a tusk of ivory to a Portuguese sailor in exchange for fabrics, it's likely that the tusk had already passed through the hands of several other tribal rulers, and each would have pocketed a commission. King Glass's cut, invariably, would have been higher than any other along that chain of supply. Despite the markups, the trading system remained a bargain for the Europeans, who never had to venture beyond the coast to fill their cargo holds. This system helped explain why, after so many years of sporadic trade, no white man was known to have traveled more than a short distance inland. To do so simply wasn't necessary, and it would have been considered a foolish risk. It would also have invited the kind of suspicion that Paul now faced among the Mpongwe: Was he simply trying to cut out the middle-man by acquiring goods directly, and more cheaply, from the inland tribes?

Most of the Mpongwe didn't trust his plan, which made his preparations more difficult than he had expected. Only his closest native friends from his years with Wilson would agree to help him acquire provisions. Even some of those who assisted him tried to talk him out of a plan they believed was unwise, if not suicidal. The jungle was no place for a foreigner, they said. The Mpongwe themselves didn't dare head inland unless it was an absolute necessity. Even then, they didn't go far.

But in his limited encounters with tribes of the near interior, Paul had learned that most of them had been eager to meet and speak with him simply because he was the son of a white trader. Among most of the inland tribes, a non-native visitor was an unheard-of novelty, and he could use that to his advantage. His status as an outsider alone was enough to grant him an audience before the king of any of the inland tribes. And if the kings were on his side, he'd never have a shortage of helpers during his journey.

The idea that so much might depend on his links to a non-African origin was a fact of life that Paul wasn't above exploiting, even though he recognized how thin and superficial such differences often seemed. Sometimes the Gabonese he met marveled at his light complexion, but Paul didn't see such a clear line of separation. "It is really tanned a very dark brown by now," he wrote of his skin after spending several weeks under the equatorial sun.

Such fine points of shading didn't matter much in the black-and-white world of western Africa in the early days of its colonization. Even if Paul seemed eager to obliterate his past for reasons he chose to keep to himself, everyone knew he was the son of a white man. That fact, both on the coast and in the interior, was much more than skin-deep.

NINETEENTH-CENTURY explorations into Africa's unmapped regions were never taken lightly. Most everything associated with them, in fact, was incredibly heavy.

Thousands of yards of cloth bundled into seventy-pound bales, bags of glass beads that weighed more than fifty pounds each, lengths of wire rolled into sixty-pound coils—and those were just the "gifts" that served as traveler's insurance, used to buy favor from whatever tribes an explorer might encounter. The standard cargo was even bulkier: an armory of guns, shields, and swords, plus hundreds of pounds of gunpowder; a full line of surveying equipment; tents, tables, and bedding; multiple kitchenware sets; crates of tea and medicine; and even a small carpentry shop for inevitable repairs. When Captain Richard Francis Burton and John Hanning Speke journeyed into Africa from the east coast in search of the source of the Nile at the same time Paul was in Gabon, their list of provisions was staggering: 380 pounds of lead bullets; twenty thousand copper caps; two thousand fishhooks; a small library of native grammar manuals and almanacs; five dozen bottles of brandy; and a trove of other items that ranged from cork beds to rain gauges. A few years later, when Henry Morton Stanley commenced his legendary expedition to find David Livingstone, he estimated that his traveling kit weighed about eleven thousand pounds. Wheeled transport was impractical when traversing such rugged terrain, and draft animals were generally useless because of the ravenous tsetse flies. As a result, a small army of native porters—between 100 and 160 men, in Stanley's case—traveled right alongside every established Victorian explorer.

Paul, alas, wasn't an established Victorian explorer.

The academy in Philadelphia hadn't provided funding up front, so his was destined to be a poor man's operation. Even so, what he lacked in provisions and equipment he made up for in ambition. His plan was to enter Gabon's interior via the Muni River, ascending it to its headwaters. Then he'd cross the Crystal Mountains on foot. In addition to encountering gorillas, he hoped to verify the course of the Congo River, which was rumored to flow northward behind the mountains. The territory was an empty white spot on the map. He hoped to color it in by himself.

Just before embarking on the trip, Paul traveled to the small island of Corisco, which sits in the bay about eighteen miles from the mouth of the Muni River. There, he would finish outfitting his expedition by hiring a few native traders to accompany him on the first leg of his journey. One of those traders was a man named Mbango, who reputedly had traveled upriver before and had bartered with a great king who ruled several tribes of the Muni. If Mbango could just introduce him to that king, Paul believed he might be able to avoid one of the most serious dangers he faced: hostilities with isolated forest tribes.

Mbango and a dozen of his friends agreed to accompany him for a couple of weeks and to lend him a canoe. The vessel was large, having been carved out of a broad tree trunk, and it was even rigged with a primitive sail. The canoe measured thirty-five feet long, three feet wide, and about three and a half feet deep.

Paul packed the canoe with everything he thought he'd need for this part of his expedition: chests containing a hundred fathoms of cloth, nineteen pounds of beads, a few looking glasses, some flints, a little tobacco, eighty pounds of shot and bullets, twenty-five pounds of powder, some basic medicines, half a dozen crackers in case of stomach upset, ten pounds of arsenic for preserving animal specimens, and his rifles.

Everything he brought with him into the jungle, including Mbango and the other men, fit into the single dugout canoe.

PAUL AND his makeshift crew, each carrying a gun, sailed across the Bay of Corisco toward the headwaters of the Muni River, which today forms part of the border between Gabon and its neighbor to the north, Equatorial Guinea. The men intended to accompany him only as far as the village of the river king, about forty-five miles inland. If Paul wanted a permanent crew to travel with him farther, he'd need to hire them in that village. These men in his canoe were traders, not explorers. The fact became painfully apparent to Paul even before they reached the mouth of the Muni.

In the middle of the bay, something caught Mbango's eye. Across the water, another boat was approaching theirs. But when the men in that boat spotted Mbango's canoe, they changed course and veered away in the opposite direction. But they hadn't acted quickly enough. Mbango recognized the boat, which belonged to a man who owed him money. Seizing the opportunity to collect, Mbango yelled to his men to pursue the fleeing men.

"But the more he called 'stop,' the harder they paddled off," Paul wrote. "Now our side became excited. Mbango called that he would fire upon them. This only frightened them more."

Mbango's canoe soon overtook the other boat. Paul's pleas for peace were drowned out in a wash of angry shouting and splashing water. Hand-to-hand combat broke out as the canoe ranged alongside the boat.

The canoe rocked violently, threatening to dump the men and all of Paul's supplies. But Mbango and his men seemed to be getting the best of their enemies. A couple of the men in the other boat jumped overboard, forcing the rest to surrender. Mbango seized three members of the other boat's crew, ushering them into his crowded canoe as captives. He told Paul that now they would make an unplanned detour: the prisoners could be detained on a nearby island, which would help guarantee Mbango the speedy repayment of the debt.

"A good day's work," Mbango told Paul.

The trader smiled at his good fortune, but Paul couldn't muster anything resembling joy. This wasn't the way expeditions into the Dark Continent were supposed to begin. Other African explorers lorded over their porters like commanding generals. Paul was comparatively powerless.

As he sailed along in the canoe, soaking wet and crammed alongside the newly procured prisoners, the young adventurer felt as if *he* were the captive.

THE CANOE slipped deeper into the Muni River, and swampy woodlands pressed closer upon both sides. The breeze was too feeble to fill the sail, so with oars the men muscled the boat up the choked tributaries that led to the king's village. Stiff mangrove roots raked the sides of the canoe. Morning passed to afternoon. They were one degree north of the equator, and at that latitude twilight arrived without fanfare: there was light, there was darkness, and very little in between.

Sounds of the onrushing evening filled the boat: the whine of mosquitoes, the slap of palms on exposed skin. Threading their way upriver by moonlight, they continued to paddle only because there was nowhere to stop and make camp. The riverbanks were muddy quagmires.

It was ten o'clock at night when they finally arrived at the village. From the canoe, they saw men, women, and children darting in and out of dancing firelight. All Paul wanted to do was meet the king, get his blessing, and collapse into sleep. But the king, whose name was Dayoko, announced that he would speak with him first thing in the morning. Paul was served a hot meal of plantains and shown to a bed. Shortly after midnight, he covered himself with mosquito netting and tried to sleep.

Dayoko was nearly seventy years old—an unusually ripe age for an African living in a village of mud and bamboo. But little else seemed to distinguish him from the others in his tribe. His meager hut lacked even the slightest trace of regal distinction. However, Dayoko had amassed an enormous collection of wives over the years. In every village for miles, he had fathers-in-law who could be counted on as allies and trading partners. Those relationships, more than anything else, gradually allowed him to accumulate more power and influence than anyone else in his tribe. If Paul could get Dayoko's blessing, he would snag a golden ticket into an invaluable tribal network that extended deep into the forest in all directions.

Shortly after sunrise, Dayoko was ready for their meeting. Paul pulled out the coat that he'd worn in New York the previous winter and presented it to the king, knowing that the exotic garment would be kindly received. But Paul didn't stop there. He offered twenty yards of cloth, a little gunpowder, some gunflints, and a few looking glasses for Dayoko's wives.

The king was duly impressed with the young man's etiquette but wasn't sold on his travel plan. Paul hoped to venture about 150 miles inland—a place that very few natives of this region had ever explored. Dayoko's sphere of influence didn't stretch that far. It was said that ferocious cannibal warriors ruled that territory. Dayoko couldn't guarantee his safety.

"Dayoko thought my project impossible," Paul wrote after the meeting. "I would die on the way, and he should have my death on his soul—a consideration which seemed to affect him greatly. I should be murdered by the cannibals and eaten. There was war on the river, and the tribes would not let me pass. The country was sick. And so on."

These were not the hysterical exhortations of a fainthearted man. Exploring the African interior in the nineteenth century was dangerous business, and Dayoko knew that his blessing would do little to reverse that fact. African explorers of Paul's day not only flirted with death; they practically invited it upon themselves. Britain sent four major expeditions to the Congo, Zambezi, and Niger Rivers between 1816 and 1841, and of the hundreds of men who started out on those journeys, a full 60 percent of them didn't make it out alive. Bare statistics, however, couldn't fully convey the endlessly creative ways that Africa tormented adventurers. Mungo Park, the first European to explore the Congo, drowned trying to swim to safety as hostile tribesmen strafed the water around him with arrows. Around the same time as Paul began this journey, Richard Burton and John Hanning Speke were preparing to venture into Somalia. In Zanzibar, they were told that the last European to attempt the feat had been a French naval officer whose luck couldn't match his curiosity: he stumbled across a group of tribesmen who tied him to a tree and systematically lopped off his limbs, one by one, ending with his head.

But even the minority who *survived* African expeditions, as Burton and Speke would, often emerged as shells of their former selves. At intervals during their journeys between 1855 and 1859, Burton took a spear through his cheek, suffered fever-induced delirium, was unable to speak because of a swollen tongue, and suffered briefly from full-body paralysis. Speke fared even worse. He went temporarily, and mysteriously, blind. As he struggled to regain his vision in a hostile environment, his tent was swarmed by tiny black beetles, one of which tunneled deep into his ear canal. Speke reported that when the insect began to "dig violently away at my tympanum," all self-control abandoned him. He desperately thrust a penknife into his ear, killing the insect but slashing apart the delicate membranes inside. The resulting infection, which eventually hardened into an abscess, inflated his face with swelling. Chewing food became impossible. "For many months the tumour made me almost deaf, and ate a hole between the ear and the nose, so that when I blew it, my ear whistled so audibly that those who heard it laughed." To make matters much worse, Speke's crew of porters—the very men to whom he'd entrusted his life— actively despised him.

But Paul remained determined, and he wasn't about to turn back after coming this far. After two days of pleading, he finally won the king's support. With Dayoko's endorsement, several members of a neighboring tribe called the Mbond-emo agreed to serve as Paul's assistants. Additionally, the king promised to send two of his own sons to accompany him.

With the arrangements made, Mbango and the rest of his men returned to Corisco, taking the canoe with them. Paul remained, alone with Dayoko's tribe. His fate was in their hands.

THE MBONDEMO had to finish their seasonal planting before they could accompany him, so he was forced to wait for a full month in Dayoko's village.

Paul had been primed to be wary of the inland tribes, thanks to years of exposure to the coastal Gabonese. Although he would make friends with members of nearly every tribe he encountered during his journeys, Paul's view of the native populations often reverted to the familiar stereotype that was a by-product of fear: the tribes of the interior were godless savages who lived by a code that was amoral and, as a result, inscrutable.

In Dayoko's village, he temporarily slipped into this mind-set. A few weeks after he'd arrived, the village buzzed with rumors of an impending execution. An old man had been accused of wizardry, of casting fatal spells upon one of the village chiefs. To Paul the man seemed harmless; bent and wrinkled, with white hair, he had been bound in crude stocks outside a hut. Paul figured he was probably the victim of tribal indifference: he was simply so old that the village didn't want to take care of him any longer. Whatever the case, the villagers had conclusively voted to kill the old man, reasoning that he'd already murdered one person with his wicked spells and was therefore a threat to everyone.

"No one would tell me how he was to be killed," Paul later wrote, "and they proposed to defer the execution till my departure, which I was, to tell the truth, rather glad of."

But before he left, he heard shrill cries rise from the river. He saw men walking through the village with blood on their hands and arms. He was told that the old man had been tied to a log and hacked into pieces.

It's impossible to know for sure if this actually occurred, but Paul gave every indication in his writings that he believed it had. On the same night as the alleged execution, the men of the village were pleasant and jocular—"mild as lambs," he wrote, as if nothing had happened. *Had they been part of the slaughter?* The possibility, in light of their absence of remorse, rattled Paul. Whether it was his imagination or not, the entire village felt sinister and unclean.

The following Sunday—a day before he was to depart with the Mbondemo for the interior—he opened his Bible, hoping to find solace. When two villagers asked him what he was doing, he tried to explain that he was reading a book that was given to the world by God—the single divine Creator who ruled all.

"Oh yes," one of the men replied, "that is true for you. But white man's God is not our God."

CHAPTER 8

"Unfriends"

London

John Edward Gray couldn't stand Richard Owen. He couldn't have put a finger on just one thing that rubbed him the wrong way; he simply hated everything that Owen represented.

In 1856, at age fifty-six, Gray was the keeper of the zoological collection at the British Museum and the vice president of the Zoological Society of London. But he had never intended to be a zoologist. As a young man, he had wanted to be a botanist, like his father before him. The story behind his failure to achieve that dream explained much about the ingrained bitterness that seemed to define his personality and his hostility toward the cultural elite.

Like any up-and-coming botanist, Gray had tried to join the Linnean Society when he was just twenty-two years old. But his membership was denied. The reason, so it was said, was that Gray had disrespected the founder and president of the group, James Edward Smith, by failing to credit him as the author of a work he had cited in a scholarly paper. Such a faux pas might have been forgiven in a layman, but not in someone vying for a place in Smith's own fiefdom. What undoubtedly made the matter worse was the undisguised contempt that Gray showed for Carl Linnaeus, the eighteenth-century Swedish botanist who conceived of the system used to classify plants and animals. Gray and his father co-wrote a paper that described the Linnaean system, which defined plant species according to their sexual parts, among other things, as the product of its creator's "prurient mind." That conclusion could not have pleased the members of the society that proudly bore the Swede's name.

After being rejected for membership by the Linnean Society, Gray shifted his focus from plants to animals. He systematically clawed his way to the top of the field, becoming the head of the British Museum's zoological department in 1840. His rise to prominence occurred *despite* the Victorian scientific establishment. And he never forgot the slight he'd received as a young man. His obituary in the *Annals and Magazine of Natural History*, a publication Gray had edited for years, tried to dissect his cantankerous streak: "One can easily understand that the circumstance of being thus ignominiously rejected must have been a bitter disappointment to a young and enthusiastic naturalist such as Gray then was; and we cannot wonder that he placed himself in decided antagonism to those whom he thought his enemies in the matter, and thus acquired that combative habit of mind which undoubtedly in after life procured him many 'unfriends.'"

His relationship with Darwin illustrates Gray's talent for stirring professional rancor. In 1848, Darwin was immersed in researching barnacles, noticing all sorts of interesting homologies between those invertebrates and other crustaceans. Gray had encouraged Darwin with his research, and he even allowed him free access to the British Museum's barnacle collections. But then, while busy putting all his research together for publication, Darwin was told that Gray "intended to anticipate" his work by publishing descriptions of the most interesting specimens himself. It was a stunning betrayal, and Darwin confronted Gray in person. "I felt anxious to know what you intended doing," Darwin later reiterated in a letter to Gray, "and I think you will admit that it was natural that I should wish that what little of novelty there yet remained in the subject, should be the reward of my work, which I assure you has been to my utmost every day." Gray backed down and continued to support Darwin, who politely promised to hold no grudges and apologized for his protestations. But the incident was awkward for both of them.

Gray's relations with Owen involved none of that politesse: they openly hated each other and made no apologies about it.

In the early 1850s, Owen and his family moved out of their house at the Royal College of Surgeons and into an estate called Sheen Lodge, which sat in the middle of London's Richmond Park. Surrounded by lush gardens, carp-filled ponds, and roaming deer, the house was a gift from none other than Queen Victoria herself. This act of royal charity caused many eyes to roll among the scientists who were forced to compete with Owen for attention and prestige, and no one was more disgusted than Gray. But in 1856, Owen was given another gift that infuriated Gray far more: he was named the superintendent of all the natural history collections at the British Museum. Owen was now Gray's boss.

Until then, Gray had ruled the department as his own personal kingdom. He had acquired a world-class collection of specimens for the museum. He lobbied hard for a new facility devoted solely to natural history, believing he should be the person to preside over it.

Owen, coincidentally, had floated the same idea around the same time. But unlike Gray, his access to the highest levels of the British government eventually enabled Owen to successfully land a new, five-acre museum facility dedicated to zoology. Naturally, Owen thought *he*, and not Gray, would be the best choice to oversee its establishment.

From the moment Owen took over at the British Museum, Gray's role in museum affairs was seriously threatened. Given the zoologist's infamous irritability, scientific London braced itself for a thunderous clash of wills. One observer who knew them both—a young professor of physiology named Thomas Henry Huxley—predicted that the two men would rip each other apart. Huxley wrote to a friend: "In a year or two, the total result will be a caudal vertebra of each remaining after the manner of the Kilkenny cats."

··◦⫞⫟◦··

CHAPTER 9

Fever Dreams

Gabon

One of Paul's most precious possessions was his medicine kit, a small but potent collection of remedies that he believed would save him from one of the deadliest evils in the jungle: a malign specter known simply as the fever.

The land is sick: he'd been fed this warning since he left the coast, and he believed it. The river near Dayoko's village was shallow, and when Paul waded through the muck, it reeked of plague. One could imagine the sickness seeping up from the ground in a wavy haze of vapors, oozing into the nostrils, settling in the rhinal cavities, doing who knew what sort of damage to a man's constitution. More than deadly beasts and warring tribesmen, *this* was what killed explorers.

The African fever of Paul's day was the same malaria that's still killing millions of people, and it was transmitted by the same kinds of mosquitoes then as now. But no one knew that. The mosquito-malaria link wasn't discovered until the 1880s. Instead, almost everyone subscribed to a "miasma theory"—the same hunch that led most Victorian scientists to believe that cholera was caused by tainted vapors and not water infected with human waste. An English traveler, venturing into Gabon several years after Paul, wrote confidently that he'd consulted "all the authorities on this subject" and collected an "immense mass of evidence" to demystify malaria's origins. He concluded that "bains, dews, winds blowing from malarious localities, marsh exhalations, and possibly the human breath, may therefore be considered as the proximate causes of fever. Those of nervous temperament, of light hair, and of fair complexion, of strumous habits, or a plethoric disposition, are the most liable to suffer from fever."

Even if the underlying cause was buried in murky conjecture, one partially effective treatment for malaria had already been identified: quinine. Dr. David Livingstone, the celebrated medical missionary–cum-explorer, attributed his relative immunity to malaria to the alkaloid, which is found in the bark of the cinchona tree. Aware of this, Du Chaillu loaded his medicine chest with all the quinine he could get his hands on. He used it not only as a remedy for fever but also as a prophylactic to prevent its onset. In the mid-nineteenth century, 2 grains of quinine per day was considered an adequate prophylactic dose. But Paul at times took 150 grains per day. He regularly supplemented the quinine with a small amount of brandy, and sometimes he took laudanum—an opiate that was a favorite of Victorian doctors, prescribed for everything from diarrhea to cholera.

Sometimes, when nothing else worked, he improvised. "And when the system becomes accustomed to quinine, and this medicine ceases to operate, sometimes a small dose of Fowler's solution of arsenic will be found very successful in stopping the chills," Paul wrote.

He occasionally blamed the quinine for rattling his nerves, but the poisons filtering through his system were probably altering his perceptions in ways he couldn't appreciate. In recent years, medical researchers have determined that quinine in very high doses can significantly lower a person's serotonin levels. Shortages of this brain chemical have been linked to depression, obsessive behavior, and even intense spiritual experiences. The anthropologist and historian Johannes Fabian, in a recent study of nineteenth-century Belgian and German explorers in central Africa, concluded that the volatile interaction of drugs such as quinine, laudanum, and alcohol undermined the rationality of the explorers, producing in them distorted states of perception that he labeled "ecstasis." Their descriptions of Africa and African natives were sensational, Fabian argues, due in part to the fact that fatigued explorers were under the influence of medications that doctors didn't fully understand.

PAUL HAD already begun regularly taking quinine before he set out on August 18, 1856, to meet the new escorts Dayoko had lined up for him. They slogged across miles of rooty thickets, through dangling vines and aloe. By the time Paul reached the Mbondemo tribe, his blue button-down shirt was torn to ribbons in places, and the skin underneath raked red.

The Mbondemo lived in an *olako*, a tribal word meaning something close to "temporary settlement." It *aspired* to be a village, but it wasn't one yet. The people slept under rectangular shelters of thatched leaves supported by four crooked sticks at each corner. After dark, when Paul arrived, the place dripped with a romantic jungle atmosphere that he loved: the orange glow of the fires lit the faces of the families and made the shadows in the forest dance. The tribe gave him, as an honored guest, one of the best shelters, with the tightest weave of leaves overhead. But the roof was overmatched by the rain, which was just beginning to regularly soak the region at night. Every morning he woke covered in mosquito bites and soaking wet.

Protection, not comfort, was what he really needed from the tribesmen. He still didn't fully trust his new companions, but as he got to know them better, his biases began to soften. Their king, named Mbene, seemed noble. Instead of invariably seeing them as godless savages, Paul began to see fellow men worthy of respect. He confided to his journal:

> Today {August 20} I sent back Dayoko's men, and am now in Mbene's power and at his mercy. He is a very good fellow, and I feel myself quite safe among his rough but kindly people. I have found it the best way to trust the people I travel among. They seem to take it as a compliment, and they are proud to have a white man among them. Even if a chief were inclined to murder, it would not be profitable in such a case, for the exhibition of his white visitor among the neighboring tribes does more to give him respect and prestige than his murder would.

Dayoko's men had promised to return to look for Paul in three months. That would give Paul enough time to explore the two ridges of the Crystal Mountains that were visible from the encampment. Those hills, looming and shrouded, had been his targets since he'd arrived in Gabon. They appeared fuzzed with light green baize, but what lurked within that verdure was the stuff of legend. Cannibals who roasted their enemies over open fires. Gorillas that made the cannibals seem harmless. He was sure to see things no one else in the outside world had ever laid eyes upon. The promise of such wonders set his fevered imagination ablaze.

CLIMBING WAS hard and slippery work. These were days of sore legs and empty stomachs, of boot soles skidding across wet stones. His barefoot companions navigated the dangers with fewer spills, but it wasn't easy for them, either. In addition to the native hunters and guides, six Mbondemo women had been assigned by male tribal leaders to serve as Paul's carriers—a lowly task that none of the men agreed to do.

All of them were rewarded with sights of otherworldly splendor. From a cliff on the side of a mile-high mountain, the valleys looked as pristine as they were mysterious. But images of Edenic solitude were broken by unwelcome visions reminding them that nature would never leave them alone.

During that first foray into the Crystal Mountains, Paul sat under a tree and spotted an enormous snake stretching across the branches overhead. He shot it, and it fell to the ground, shook reflexively a few times, and died. It measured just over thirteen feet long. To Paul, it was a repulsive creature, and it made his skin crawl. To his companions, it was lunch. They flayed it, roasted the meat over a fire, and ate it. Paul couldn't stomach the sight of it. But it wasn't as bad as the mangabey monkey the hunters cooked days later. That image rattled Paul to the bone. He thought it looked like a roasted baby.

The more time he spent with his guides, the more his perceptions grew into fantastic shapes, colored by fear, flowering into surreally vivid, almost psychedelic, phantasmagoria. His hatred of snakes soon populated the jungle with terrifying serpents that were eager to unhinge their jaws and engulf him. Once, when he was stepping into a cave with his companions, his mind immediately conjured a squirming mass of snakes—a fantasy, but one that reality nearly trumped. "Peering into the darkness, I thought I saw two bright sparks or coals of eyes gleaming savagely at us," he wrote. "Without thinking of the consequences, I leveled my gun at the shining objects and fired. The report for a moment deafened us. Then came a redoubled rush of the great hideous bats; it seemed to me millions on millions of these animals suddenly launched out on us from all parts of the surrounding gloom; our torches were extinguished in an instant, and panic-struck, we all made for the cavern's mouth—I with visions of enraged snakes springing after and trying to catch up with me."

At the very forefront of his imagination, however, Paul reserved a special place for the King of the Forest—the gorilla, for whom he always remained vigilant, wherever he went.

WHEN PAUL saw the strange footprints near a cluster of sugarcane where a native village had once stood, the native hunters who had accompanied him all day on the long, slippery uphill slog said one word: *nguyla*. Paul recognized it as the Mbondemo version of *njena*, or "gorilla."

As they followed the tracks to large granite boulders, they saw that the tracks included knuckle marks, suggesting the animals had been walking on all fours, stopping occasionally, it seemed, to sit and chew stalks of cane that they had taken from the patch. Examining the scene closely, the men counted at least five distinct sets of prints. They seemed fresh.

When Paul and the men got their first fleeting glimpse of the gorillas, the tree cover was too dense to get off a clear shot. But the momentary flash of their bodies—Paul thought he might have seen four different gorillas run for cover—was vividly impressed in his mind.

Considering that they had been climbing the mountain-side since morning, they should have been exhausted when they returned to the camp that night. But they didn't sleep. Instead, they replayed the scenes from that day, soaking their impressions of the sightings in the volatile fuel of folktale and legend.

Paul had been hearing the natives' stories about gorillas for years—about how they were more powerful than lions, about how they could kill men with brutal efficiency. But around the fire, his guides took those stories to another level. They lit up the night with descriptions of a beast that bore only a passing resemblance to the skittish phantoms they'd barely glimpsed hours before.

One of the men told Paul a story—a *true* story, he assured him—about two of the women from the Mbondemo tribe who had encountered an enormous gorilla deep in the forest. The beast seized one of the women and disappeared with her into the gloom. The other woman ran back to the village in hysterics. But the tribe could do nothing to help, and they mourned the victim's certain death. A few days later, they were shocked when the abducted woman returned. The gorilla, she reported, had raped her, leaving her traumatized but otherwise unharmed.

"Yes," one of the Mbondemo men told Paul, "that was a gorilla inhabited by a spirit."

Stories of gorillas that were possessed by humans, Paul learned, constituted an entire chapter of the native lore. Possessed gorillas could be distinguished, they said, by their unusual enormity and the fact that they could never be killed. These men-beasts were the perfect union of brute strength and human intelligence, they said. The men rattled off names of members of their own tribe, now dead, who were believed to roam the forests in the bodies of gorillas.

Paul didn't take these stories as literal truth, but where could he draw a line separating reality and tall tales when it came to less supernatural matters? He'd heard more than one tribe say that gorillas were known to lurk in trees, waiting to ambush people who walked underneath, pulling their victims up with their feet before quietly choking them with their massive hands. Some said that the brute exercised a bloody dominion over all the animals in the forest. Most of the native tribes agreed that gorillas sometimes beat elephants to death with sticks.

These men, reliable or not, had more knowledge concerning gorilla behavior than anyone else alive. No one else, anywhere in the world, had as much experience with the animal as they had—even if they'd be the first to admit that they always tried to *avoid* accumulating such experience if at all possible.

But now, after the brief encounter of that day, Paul was an expert. He'd gotten one little glimpse of gorillas in their natural habitat, and that was enough to make him the world's foremost authority on the behavior of the most fantastic beast in the annals of natural history. No one else, aside from a handful of native tribesmen, had ever seen this much. He was desperate for more.

·⊰][⊱·

CHAPTER 10

Between Men and Apes

London

When T. H. Huxley predicted that John Gray and Richard Owen would rip each other apart if forced to work together, it had seemed like a real possibility—but only if Huxley himself didn't get to Owen first.

Like most up-and-coming scientists in Victorian London, Huxley had never really liked Owen. As early as 1851, when he was just twenty-six, Huxley observed in a letter to a British entomologist: "It is astonishing with what an intense feeling of hatred Owen is regarded by the majority of his contemporaries." Although Owen hadn't wronged him personally, Huxley thought the older scientist arrogant and egotistical—"a man with whom I feel it necessary to be always on my guard."

Now, in 1857, Huxley found his opportunity to try to knock Owen off his lofty perch.

Early in the year, Owen delivered a lecture to the Linnean Society arguing that the brain of man differed from that of the gorilla in a fundamental structural way and that the differences could never be bridged by transmutation. Specifically, Owen cited three structures present in the architecture of man's brain that he said didn't exist in apes: the posterior lobe, the posterior cornu, and the hippocampus minor (a spur on the horn of the brain's lateral ventricle that today is known as the calcar avis). Owen argued that the structures allowed the development of larger and more powerful brains, and as a result humans deserved a taxonomical upgrade. Not only did men deserve to be classified in a separate order from apes, Owen argued, but they should be classified in a completely different subclass.

Owen framed the discussion as pertaining not just to physical science but also to metaphysics. The particular biological characteristics that distinguished man from all other animals were precisely the same ones that allowed man to fulfill "his destiny as the supreme master of this earth, and of the lower Creation."

The *London Lancet*, the journal of Britain's medical establishment, strongly endorsed Owen's writings on gorillas and the human brain. By methodically emphasizing the differences between man and ape, Owen would become, in the words of the journal, nothing less than a vindicator of "the dignity of the human race."

Huxley, then a thirty-two-year-old professor at the Royal Institution of Great Britain, wasn't convinced. He began dissecting Owen's Linnean Society lecture, searching for flaws.

"As these statements did not agree with the opinions I had formed," Huxley later wrote, "I set to work to reinvestigate the subject, and soon satisfied myself that the structures in question were not peculiar to Man, but were shared by him with all the higher and many of the lower apes."

What, then, was the defining difference between a man and an ape? The question had never seemed so urgent.

··⊰[]⊱··

CHAPTER 11

Maps and Legends

Africa

Ignoring everything green, Paul shot the color out of the forest. The gray and red of the African parrots. The scarlet of rosy bee-eaters. Yellow weavers, purple herons, indigo swallows. The black-and-white palm nut vultures.

"The dry season is delightful in Africa," Paul noted in his journals. "It is the season of flowers, of humming-birds—who flit through bushes at all hours, and charm one with their meteor-like flight—of everything pleasant."

Life had settled into something like a routine by the middle of 1856. He'd wake in the morning at about five o'clock and down a cup of strong coffee. Then he'd grab a rifle and hunt birds until ten. After that, he'd return to his encampment for a starchy breakfast of plantains and manioc. He filled out the remainder of the morning stuffing his fresh bird specimens with cotton and carefully curing the skins in a solution of arsenic to protect them from insect-driven decay. He'd rest until mid-afternoon, when he'd hunt for about three more hours. Dinner was served around six. The rest of the evening was reserved for more small-scale taxidermy and chats with the tribesmen.

Following this schedule, Paul earned a reputation among all the tribes he encountered as more of a hunter than an explorer. Shooting was the one part of his life where his pride showed through strongest. To him, hunting wasn't mere sport. It provided a foot in the door of a scientific community to which he didn't yet belong.

Today science has uncoupled itself from hunting, but the two realms were indivisible for most of the nineteenth century. A naturalist, more often than not, had to kill in order to thrive. This was viewed not as crude or barbaric but as a prerequisite of developing a heightened sense of appreciation for the natural world. Before someone like John James Audubon could hone his gift for revivifying birds on an artist's canvas, he had to develop his skills as a marksman. In those days before field photography and telephoto lenses, stuffing and wiring animals into lifelike poses was considered the best way to enable the sustained reflection required to deepen one's respect for nature. Audubon himself shot most of the birds he drew—a necessary compromise that inspired generations to refine their own appreciations of nature and, in some cases, to work to protect such species from endangerment. In the twentieth century, that compromise became unnecessary. Hunting lost whatever scientific, academic, and artistic authority it once claimed. But in the mid-nineteenth century, naturalists felt little moral pressure pushing them away from hunting. If anything, they were pushed *toward* it by scientists and academic institutions with a wolfish demand for specimens.

No one exhibited a stronger desire for bird specimens than John Cassin, Paul's contact at the Philadelphia academy. Cassin sponsored a small army of guns for hire who spanned the globe, shooting, stuffing, and shipping birds across the seas to his museum in Philadelphia. He had catalogued and named nearly two hundred different species of previously unidentified American birds. By the mid-nineteenth century, from his perch in Pennsylvania, he was working his way around the world, building new taxonomies for birds native to Asia, Africa, and South America. Cassin had reached the pinnacle of his field and had earned respect on both sides of the Atlantic.

A young naturalist eager to make a name for himself would have had a hard time finding a better model than Cassin, and Paul diligently supplied him with stuffed specimens. When he was close to the coast, he'd ship them aboard vessels bound for America, the birds laid out stiffly on their backs, plump with cotton stuffing. Within months, Cassin was standing in front of the other members of the Philadelphia academy, telling them that the young adventurer who'd so recently begun his journeys had already shipped him more than one thousand specimens.

One of the previously unclassified birds that Paul had packed in a box and shipped to Philadelphia was a small, acrobatic insectivore with a short bill, a muted brown breast, and dark wing tips. After Cassin examined it, it ended up in the hands of Ferdinand Heine, a German ornithologist who had the privilege of bestowing an honorific on the bird. Paul's bird was christened *Muscicapa cassini*, a name that it still carries today: Cassin's flycatcher.

It was clear that Paul, as an outsider to the scientific establishment, wouldn't make a name for himself even if he sent a million birds to the States.

———

AFTER HE got his first glimpse of a gorilla in the shadows of the Crystal Mountains, Paul jumped at any chance he could get to shake up his daily routine to try to find another one. But lowland gorillas generally are encountered more easily by accident than by hunting for them. Because they range widely in small groups and are very difficult to track, they're the kind of animal you stumble upon during long treks in the most remote parts of the forest, and not the kind likely to reward a hunter's pursuit. Paul and his Mbondemo helpers often spent full days simply hiking eastward, moving their portable camp farther inland. Some days he estimated they walked as far as twenty miles through the forest. They rarely traveled in a straight line.

In the middle of one of these day treks, when time was counted by the steady rhythm of footfalls, one of the men with him stopped walking and clucked his tongue—a signal to everyone to be quiet and still. They froze and peered eagerly at the forest around them. The rustle of tree branches. The rip of uprooted stalks and stems.

Paul's first thought was of gorillas. The grave expressions on the faces of the other men betrayed the same hunch. They checked their guns, making sure that they had powder in the pans of their rifles. Slowly, they tried to proceed toward the noise, stepping as lightly as they could across the undergrowth. They walked within a silence so profound that even the sound of breathing, Paul thought, seemed loud. Through a dense screen of branches and leaves they finally saw movement—something near a fruit bush ahead. Without warning, the silence was ripped apart by a husky roar.

He had never heard a sound so unsettling: a terrifying bark that seemed to start somewhere deep in the pit of the animal's massive chest and explode outward in a thunderous crescendo.

For the first time in his life, Paul got a clear view of a gorilla—an adult male with a tuft of silver hair on its back. It rose onto two legs when it saw the hunters and appeared to stand almost six feet tall. The forearms bulged with the promise of strength, its neck a massive pillar of solid muscle. The animal must have weighed nearly four hundred pounds. With the hunters drawing their guns, the gorilla charged but didn't directly attack; it stopped about six yards shy of them. Du Chaillu heard the blast of a gun and watched the gorilla fall forward onto its face, with a groan. The men watched the animal convulse for a moment, arms struggling against the dirt. Then all was still.

After years of picturing the gorilla in his imagination, Paul finally got the chance to examine one up close, but the other men in the group had different plans.

Paul measured the carcass at five feet eight inches. He noted the astonishing girth of the forearms and the deep gray of the eyes. While he worked, his men were building a fire and a crude shelter on the spot. As Paul continued to study the animal, the men had already begun to divvy up the parts of the gorilla among themselves. Paul couldn't stop them; they slaughtered the animal, roasted it, and tore into its deep red flesh. Eventually, they cleaved the head to expose the brain, because the Mbondemo had heard that consuming it offered two benefits: it strengthened a man's powers as a hunter and as a lover. They took particular care with the valuable, spongy pink tissue and shared it equally.

"Luckily, one of the fellows shot a deer just as we began to camp," Paul wrote, "and on its meat I feasted while my men ate gorilla."

The men's eagerness to eat that first gorilla was an indication of how extraordinarily rare it was for them to kill one.

The tribesmen were not great hunters, partly because they were forced to hunt with cheap, African-made flintlock muskets that were notoriously unreliable and imperfectly calibrated. The thin barrels were easily bent. Usually, the native hunters poured powder of impure quality into their guns, then stuffed dry grass on top of it as wadding. The bullets were often crude bits of old iron, stuffed snugly into the barrel with more grass. As often as not, the degraded powder exploded without successfully firing the bullet—a failure known as a flash in the pan. They customarily overloaded their guns as a result, and when the rifles *did* successfully fire, the explosions often were so volatile that the weapons' components were damaged in the process. This unpredictability led many hunters to hold the rifles in front of them, a short distance from their shoulders—a precaution that sacrificed accuracy for personal safety.

Paul, on the other hand, was equipped with comparatively sophisticated rifles, powder, and ammunition. He was also a truly skilled marksman and often amazed the native hunters with his accuracy. Even so, his abilities were certainly exaggerated by his companions, given that almost none of them had previous experience with state-of-the-art firearms and powder. They speculated that he had, in fact, eaten gorilla in the past—how else to explain his gifts as a hunter who always seemed to get the better of his prey?

As the months stretched on, the Mbondemo tribesmen used his guns and powder to become formidable hunters in their own right, securing spots for themselves in the local folklore. As more of the animals began to fall to their rifle fire, the Mbondemo hunters quickly lost the desire to eat their kill. Instead of using his arsenic to preserve bird skins, Paul began using it to preserve the carcasses of gorillas.

John Cassin back in Philadelphia could have all the birds he wanted. The gorillas were Paul's.

IN 2001, scientists split gorillas into two different species, which are very closely related and share many of the same physical and behavioral traits. Western gorillas—including the lowland subspecies that Paul encountered—make up one of the species, and they range across the forests of equatorial Africa in the Congo River basin. Eastern gorillas, first discovered in 1902, are the other species; they're found on volcanic slopes of the Virunga Mountains and in the eastern forests of the Democratic Republic of the Congo.

Paul's western lowland gorillas represent the most common subspecies; the vast majority of gorillas born in captivity in zoos and research institutions are of this variety. But they are also the least studied in the wild, mostly because they range over a much wider territory than do mountain gorillas, and that jungle terrain is generally more difficult to penetrate. Adult western lowland males weigh about twice as much as females, commonly topping out at around four hundred pounds. The adult males are called silverbacks because of the silvered hair they develop after about twelve years of age. They generally live in groups of between seven and sixteen members, which are roughly analogous to families. Commonly, these groups consist of a single silverback male, three or four adult females that mate with the male, and their resulting offspring. Life spans vary in the wild, but the oldest gorilla ever verified by researchers died at age fifty-three. Like chimpanzees, gorillas are highly intelligent.

As Paul examined more and more specimens, he confirmed something that is now a well-established fact: gorillas are, for the most part, herbivores. Every time he sliced open one of their stomachs to examine the contents, he found only vegetable matter—not exactly the stuff of bloodthirsty demons. But it was an irony of nature that those vegetarian proclivities had a direct hand in shaping the gorilla's frighteningly imposing aspect. Those powerfully massive forearms, as muscular as human thighs, are perfect for tearing rough plants out of the ground and snapping thick stalks. Those formidable teeth and powerful jaws are necessary to bite, break, and grind coarse stems and trunks, and a muscular neck is required to support the temporal muscles used to chew. An oversized torso is essential to house the enormous digestive organs needed to handle such bulky, fibrous fare.

Paul noted the shiny black skin visible only on the hairless face, hands, and chest. The ears seemed miniatures of the human form. He measured the teeth, the bones, the nasal apertures. He noted that the skin sometimes felt as thick as oxhide but, fortunately for him, wasn't so thick that a close-range bullet couldn't penetrate it with ease.

Paul began to distribute high-grade powder to everyone in his hunting party. His traveling chests started to fill with gorilla skins. Dozens of them. But his fear of the creature didn't die as easily as the animals did.

Once, when the rainy season was drying up, the hunters descended into a dark patch of forest where visibility was limited to about ten feet in any direction. It was a known haunt of gorillas, according to one of the tribesmen.

The men decided to split up, which was customary. One of the men ventured into the woods alone. Paul and a man named Gambo, the son of a local chief, veered in another direction. A couple of other hunters split off on still another route.

According to Paul's later account of the event, one gun-shot rang out in the forest, then another. He and Gambo ran toward the shots. The roar of a gorilla flooded the woods and was soon replaced by silence. After searching, they found a figure lying in blood on the forest floor. Expecting to see a gorilla, Paul was surprised to see a man—the hunter who'd ventured out on his own. He was bleeding from the abdomen, his bowels spilling out of a deep lacerating wound. His gun lay beside him. The stock was broken, and the barrel was bent.

For two days the man lay in pain in their camp. Paul, when he wasn't trying to snuff his own fevers by self-medicating with quinine, tried to nurse the wounded man back to health with brandy, the only thing in his medicine chest he could imagine doing any good. The wounded man drifted in and out of consciousness, and a story began to take shape to explain his injury. The hunter's first shot at the gorilla merely grazed the animal, it was said, and as he was frantically reloading for a second shot, the gorilla attacked him and delivered a powerful blow to the abdomen. The marks on the gun barrel seemed to be from the teeth of the silverback.

On the third day, the man died and the story entered the realm of legend. The natives decided that the gorilla that killed him had been possessed by a human—"a wicked man turned into a gorilla," they said, that "could not be killed, even by the bravest hunters."

But a day later, Paul's men returned to camp with the carcass of a male silverback. He assumed it was the same animal that the unlucky hunter had encountered, but like so many things about this jungle, he'd never really know for sure.

———

IF THE people back in the States were wowed by the stories he could tell before this trip, when his experience had been limited to the coast, just imagine how they'd react to him now.

Murderous beasts! Hissing serpents! Naked savages! Stories that had been lurking in these forests for centuries, untold, growing wilder every day. And they were all his for the taking.

CHAPTER 12

A Lion in London

Richard Owen rose from the head table and turned to address the 350 people who had packed the Freemasons' Hall on a February evening in 1858. He lifted a glass of wine to toast the man who sat to his right—"the distinguished traveler we have this day assembled to honor," Owen said.

David Livingstone probably was, at that moment, the most famous man in England. The Scottish missionary had spent years exploring Africa, and his memoir about his experiences—including his terrifying encounter with the lion that had mangled his left arm—was the hottest book in the country. Since returning from Africa, he'd spent the past year collecting honorary degrees, getting mobbed by adoring crowds at sold-out lectures, and hosting a constant parade of dignitaries. His public image couldn't have been better: Livingstone was a brave and humble adventurer who, as Owen told the audience, sought to spread "that higher wisdom which is not of this world." Just that morning Livingstone had enjoyed a private audience with Queen Victoria, who wished him luck on his latest venture: an expedition to try to open the Zambezi River to travel. The upcoming expedition—one of the most ambitious ever launched—was being sponsored by the Royal Geographical Society, which had also arranged this elegant banquet as a formal send-off.

Owen recounted to the audience how he'd met Livingstone eighteen years before, when the young missionary came to him seeking advice on collecting natural history specimens before his first trip to Africa. Since then, Livingstone had occasionally provided Owen with specimens that ranged from dinosaur bones to elephant tusks. Owen returned the favor by helping him edit his book to make sure the biological descriptions were sound. Livingstone, whose wife was sitting beside Caroline Owen up in the ladies' gallery, had become a good friend of "the professor," as he sometimes called him. Years later Livingstone would joke that his mutilated left arm, which had become an iconic symbol of bravery, should be bequeathed to Owen when he died. "That is the will of David Livingstone," he said.

Livingstone might have felt indebted to Owen for conferring scientific legitimacy on the missionary's work, which in turn had helped the missionary win the respect of learned academies all over the world. But Owen's support on this night was perhaps less a favor to Livingstone than it was to his good friend Roderick Impey Murchison, the president of the Royal Geographical Society.

Murchison had personally micromanaged almost every detail of the banquet, from the precise sequence of the toasts to the Scottish airs played by the band. Molding Livingstone's celebrity was Murchison's highest priority. Before Livingstone even knew he might want to write a book about his adventures, Murchison had lined up a publisher and secured a deal. When Livingstone was mobbed in the streets of London, it was Murchison's advance public relations work that had made it happen.

Like Owen, Murchison genuinely liked and respected Livingstone. But his advocacy wasn't without an ulterior motive.

SINCE 1843, Murchison had been elected three different times to serve as president of the Royal Geographical Society. When the institution was granted its royal status, Murchison had been the only person named in its charter as a founder. In the eyes of many, he *was* the Royal Geographical Society.

In his younger days, Murchison had lived a life of leisure as a country squire. But at the urging of his wife, he took up geology, a subject of which she was fond, and he began devoting his ample free time to its study. His inherited fortune allowed him to finance trips to Scotland, Russia, and the Alps, and soon he'd risen to the summit of the discipline. His studies of mountain formations and classifications of geological strata ranked among the most important developments in the burgeoning field.

But by the late 1850s, he was less known as a geologist than as a patron for the explorers of the RGS. Following three centuries of ocean-faring expeditions, the mid-nineteenth century had become the golden age of inland explorations, and no one could claim more responsibility for that than Murchison. To some, he appeared like an imperial chess master, moving his pawns around the world, expanding the British domain one newly explored territory at a time. His influence jumped off the pages of any world atlas: among the twenty-three topographical features on six continents that eventually bore his name were Mount Murchison in Antarctica, Murchison Falls in Uganda, Murchison Island in Canada, the Murchison River in Western

Australia, plus its two tributaries, the Roderick and the Impey. Explorers all around the world revered him, at least when it came time to dole out glory, because he was their lifeline: the man with a hand on Queen Victoria's purse.

Africa held special charm for him. Livingstone sparked an unprecedented mania among the British public for stories of African adventure, and Murchison helped sate their demand. He called his core group of Dark Continent explorers—Livingstone, Richard Burton, and John Hanning Speke, among others—his "lions," and he nurtured them with special care.

The RGS was riding the public mania for exploration for all it was worth. Membership was exploding, and its monthly meetings had become *the* place to be seen among the city's upper classes.

When Livingstone left London in 1858 for his latest expedition, which promised to last several years, Murchison had identified the key to the continued growth and funding of his empire within an empire: heroes with thrilling stories to tell.

··◦][◦··

CHAPTER 13

The Man-Eaters

Africa

The only inhabitants of that forest who rivaled gorillas in terms of mythical baggage were the members of the Fang tribe. Today they're the dominant ethnic group in Gabon, making up about 80 percent of the nation's population. But in the mid-nineteenth century, the tribe was still confined to the interior of the country. They were steadily migrating westward but hadn't yet made it far beyond the Crystal Mountains. Even so, the Fang were legendary on the coast. The tribe was said to have a taste for human flesh.

Few things excited the imaginations of European and American travelers as did rumors of man-eating tribes. Stories of cannibalism went hand in hand with newly explored territories all over the world. It didn't matter that the cannibalism itself was almost never witnessed directly. History was still full of stories about the ultimate taboo.

It started with Herodotus, who wrote that the "Andropophagi ['man-eaters'] have the most savage customs of all men; they pay no regard to justice, nor make use of any established law." That term, often spelled "anthropophagi," had a monopoly on describing the practice of eating human flesh until Christopher Columbus sailed for the New World. In his journal on November 23, 1492, Columbus reported that natives in the West Indies spoke of man-eaters who inhabited a nearby island, and they called them *caníbales* (a variable spelling was *caribes*, which provided the name to the region that Columbus explored—the Caribbean).

After Columbus, cannibalistic natives became a staple of exploration literature, described by everyone from Captain Cook to Herman Melville. The term was over-applied from the start, due to the unhealthy mixture of ethnocentric fear and a lust for conquest. Labeling a race as "cannibalistic" was as good as saying it deserved to be annihilated. Queen Isabella of Spain issued a decree in the sixteenth century stating that the only natives that Spanish colonizers could legally enslave were cannibals. If they ate human flesh, they were beyond hope, and they didn't deserve to benefit from civilized behavior. Perhaps as a consequence of this decree, rumors—often initiated by rival tribes hoping to save themselves from the onslaught of colonization—were unquestioningly accepted as fact.

But Columbus himself had pointed out an irony that was easily overlooked: the same natives who had told him about the *caníbales* had also believed that the European newcomers were man-eaters when they first encountered them.

IN GABON, the Mpongwe tribe that controlled coastal trade had a natural incentive to exaggerate the Fang's reputed ferocity: they wanted to discourage travelers from dealing directly with inland tribes. Everyone on the coast had heard stories about roving bands of Fang who offered to buy dead bodies from other tribes, just to satisfy their hunger for human flesh. But the Mbondemo tribe that Paul now traveled with had a more levelheaded view of the Fang. King Mbene maintained sporadic relations with the residents of three Fang villages, and he could speak their language. He had even taken a couple of Fang women as wives, purchasing them with ivory.

Paul's first encounter with the Fang caught him off guard. He had been in the forest, looking up in a tree at a chattering monkey, when he looked down to see a male Fang warrior and two of his wives standing silently before him. The man held two spears and a huge shield of elephant hide, more than three feet high and two feet wide. The two women stood by two large woven baskets, which they'd been carrying on their heads before they had spotted Paul. They were clothed only in the skin of a wildcat, which hung from a woven strand of soft tree fibers around their waists. Their hair, as well as the man's beard, was braided in long thin plaits, each one ending with either white beads or metal rings. One glance at them was enough to set Paul on edge. But the three Fang appeared to be just as scared as he was.

Almost immediately, Paul was joined by some Mbondemo men who escorted the Fang to their encampment. Paul offered them beads to prove his friendliness. They accepted them.

Within two days Paul and King Mbene set off to visit the Fang in their village.

Paul walked into their community like a man wading into dangerous waters, vigilant for anything suspicious that might be lurking under the surface. The village was fenced for security, and a dog barked upon their approach. A single dirt lane about eight hundred yards long was flanked by thatched-roof huts. Paul spotted the bloody remains of something on the edge of the village: the more he looked at the mess, the more *human* it began to appear to him. A woman walked by holding what appeared to be a bone—*was that a man's femur?*

The ruler of the village met him in the communal palaver house, the brass rings around his ankles jangling with each step. He wore red body paint, and his face, chest, and back were heavily tattooed with blue curving lines. When the man opened his mouth to speak, Paul thought his teeth looked uncommonly sharp, as if they'd been filed to razor-edged points.

King Mbene, greeting his Fang counterpart, seemed to instantly acquire a few extra cubits of stature, and it had everything to do with the adventurer standing at his side.

"Mbene is in great glee," Paul wrote after the encounter, "as wherever he goes he is surrounded with Fang fellows, who praise him for being the friend of white men."

The Fang prepared a house for Paul. That night he barred the door with a traveling chest so no one could enter while he slept. When the king's wives brought cooked bananas to his hut, he could summon no appetite; he imagined that the bananas might have been boiled in the same pots the Fang used to cook human flesh.

He couldn't drive the horror of cannibalism from his brain, just as he couldn't wholly suppress a simple observation that seemed to rebut their savagery: these were the *nicest* man-eating barbarians a lonely wanderer could ever hope to encounter.

Throughout his stay in the village, Paul was treated as if he were the guest of honor at a great ceremonial banquet—not the main course. The Fang ruler, a man named Ndiayai, himself took Paul out elephant hunting and proudly taught the curious visitor everything he wanted to know about the tribe's hunting methods, weaponry, agricultural techniques, and ceremonial rituals. He even introduced him to other Fang populations that were scattered in other parts of the Crystal Mountains region.

"Today, several hundred Fang from the surrounding village came in to see me," Paul wrote. "Okolo, a great king among them, gave me his knife, saying it had already killed a man. Tonight there is a great dance in honor of the arrival of a spirit (myself) among them."

When he bade farewell to the Fang, the tribe seemed truly sad to see him go, and they presented him with gifts and promises of loyalty and affection. Paul never dropped his certainty that they were cannibals. But just as the *caníbales* that Columbus had described believed the Spaniards themselves were man-eating savages, it seemed that the Fang harbored their own myths concerning Europeans.

One of them confessed to Paul that his tribe had heard stories about the fiercely cannibalistic ways of white men. Paul's first instinct was to laugh him off as a simpleminded fool. But the legend hadn't been conjured from thin air. When Paul tried to assure him that white men didn't eat black men, the man confronted him with a direct challenge: explain why they bought and sold Africans as if they were cattle, not human beings.

"Why do you come from nobody knows where, and carry off our men, and women, and children?" the man asked Paul. "Do you not fatten them in your far country and eat them?"

CHAPTER 14

D.O.A.

On September 10, 1858, a package from Gabon arrived at the Crystal Palace in London. It was a cask of rum.

Owen and a taxidermist named Abraham Bartlett pried open the cask the day after it was received. A vile stench wafted out, and they reflexively slammed the lid of the barrel shut. They moved the cask outside and opened it again, summoning the requisite willpower to squint past the fumes. Hair and skin floated loosely in the rum, detached from what appeared to be a carcass.

It was a juvenile gorilla. Natives in the interior of Gabon had carried it all the way to the coast, and someone had put the remains in the alcohol to try to save it. Unlike Paul Du Chaillu with the specimens he had been collecting months prior to this shipment's arrival, the unknown person who had found the carcass or killed the animal didn't preserve it with arsenic. It had decomposed severely.

At Owen's urging, Bartlett labored heroically for a week—in an open field, hungry for fresh air—trying to reassemble the remains of the undersized specimen. The result was a pitiful approximation of the living animal. It provided little of scientific value.

IN THE first months of 1859, Owen continued to deliver lectures stressing that the gorilla was man's nearest approximation in the animal kingdom. But some people were beginning to wonder if they'd ever have a chance to get a good look at one.

The editors of the *London Lancet* in 1859 wrote: "Whether we shall ever be treated to a sight of a living animal is a doubtful matter. There is evidently much more difficulty in obtaining a young gorilla for exhibition than a young chimpanzee; and if no full-grown chimpanzee has ever been captured, we can scarcely expect the larger and more powerful adult gorilla to be ever taken alive. It is said that a bold negro, the leader of an elephant-hunting expedition, being offered a hundred dollars if he would bring back a live gorilla, replied, 'If you gave me the weight of yonder hill in gold coins I could not do it.' "

CHAPTER 15

Spirit of the Damned

Africa

Unable to travel farther inland beyond the Crystal Mountains, Paul retreated to the coast and tried his luck about two hundred miles farther south. He sailed to Cape Lopez, where he planned to follow the course of the river system that stretched inland from the Fernan-Vaz Lagoon. Then he would hike as far as he could on foot.

As he prepared for this leg of the expedition, he spent fifty dollars to build a five-room house near the banks of the Fernan-Vaz. He stored his gunpowder and ammunition in a shack, and he kept a hundred chickens and a dozen ducks in a fowl house. He hired twenty men from the coastal Nkomi tribe to accompany him on his upcoming journey, and they stayed in the dozen tiny huts that ringed Paul's house.

He'd created his own village, essentially. With a warm sense of satisfaction he'd survey the wide prairie that stretched in front of this homestead, watching white eagles soar over sprawling groves. A stream burbled in front of the property.

"Looking upstream almost any time," he wrote, "I can see schools of hippopotami tossing and tumbling on the flats."

But a sinister traffic also occupied those waterways. The Fernan-Vaz river system was a hotbed of the slave trade, much of it illegal. About a dozen miles from Paul's new home, he found a pair of holding pens run by Portuguese slavers. These barracoons, or "slave factories," as they were called, were little more than prisons where men, women, and children were held before being sold and loaded onto ships bound for the Americas. During the 1850s, most of those ships were bound for Brazil or Cuba. Paul asked those in charge for a tour of the premises, which they willingly provided.

Even before this expedition, he opposed the institution of slavery, sharing John Leighton Wilson's objections to the shackling and confinement of any potentially divine human soul. His observations of the Cape Lopez barracoons only reinforced his opinion. The slaves milled around in holding pens outside, bound together in groups of six by chains that connected their iron collars. The slaves represented a diverse mix of tribes, some from deep in the interior, and they were bound together without regard for tribal affiliation or language. The result was that a slave more often than not couldn't speak to the others chained to him. But if he wanted to drink, for example, he had to coordinate the entire group's movements toward one of the big buckets of water that had been left out for them in the pen. Huge cauldrons sat in the corners where Portuguese overseers would cook the beans and rice fed to the slaves. Some of the captives looked abjectly miserable. Others seemed oddly serene, as if they'd made peace with their fates.

When he visited the second factory, which was much like the first, he saw a boy of fourteen sold for a twenty-gallon cask of rum, a few fathoms of cloth, and some beads. But that sale was just the first of dozens that day, because a Brazilian slave schooner was anchored near the shore. Paul watched for two hours as six hundred slaves, chained together in groups of six, marched from the factory to the shore. The men and women filed into enormous canoes, which were powered by no fewer than twenty-six paddles, and then were loaded into the schooner's narrow hold.

"They seemed terrified out of their senses; even those whom I had seen in the factory to be contented and happy, were now gazing about with such mortal terror in their looks as one neither sees nor feels very often in life," Paul wrote.

He soon discovered that the peaceful grove nearest his house hid dark secrets. While hunting for birds, he heard the clanking procession of a dozen chained slaves carrying the corpse of another. A Portuguese overseer marched behind them, holding a whip. Paul followed them to the edge of the grove. They threw the body on the bare ground, then marched off in the direction of the factory.

He waited until they left to investigate the grounds, which had already attracted the notice of vultures circling above. As he walked toward the body, he noticed that the ground was littered with old bones.

"The place had been used for many years, and the mortality in the barracoons is sometimes frightful," he wrote. "Here was the place where, when years ago Cape Lopez was one of the great slave-markets on the west coast, and barracoons were more numerous than now, the poor dead were thrown one upon another, till even the mouldering bones remained in high piles, as monuments of the nefarious traffic."

The inescapable reality of the slave traffic pervaded everything, even the relationships among the free men Paul employed for his expedition. He knew that many of his acquaintances back in the United States—who lumped all native Africans in the same degraded social class—would have struggled to comprehend the subtleties of the African social hierarchy. Most of the coastal tribes, for example, considered themselves vastly superior to the tribes of the interior, whom they themselves often kept as slaves. His men in Cape Lopez told him that the stigma of slavery was a tough thing to shake. If a free man and a slave had a child together, for example, the child was considered free, because it was a paternal society. But the child's freedom was a technicality. His social rank was forever tainted by the mother's degraded position. "Even in this rude Cape Lopez country to be born of a slave mother is a disgrace, and debars the unfortunate from much of the respect and authority which his daily companions enjoy," he discovered.

Paul returned to the village he'd created as if it existed in a sphere separate from all of that. His was a place where the past couldn't shadow a man, where relationships beyond a person's control didn't dictate status, where parentage wasn't fate, and where all slates were clean. His village was the capital of a new, self-made universe where Paul was able to redefine himself. Here, he could be an American, a brave explorer, a slayer of beasts. In this world of his own creation, he had discovered the freedom to be anything he wanted to be.

He named his village Washington.

EXPLORATION, LIKE war, is usually a long succession of unremarkable moments punctuated by brief flashes of action. Those flashes, though not in the least bit representative, tend to define the experience. When Paul and his men eventually pushed into the interior again, they endured months of arduous slogging through the overgrown forests. But the miserable routine was made bearable by a few radiant moments of remarkable incident.

On May 4, 1857, Paul's hunters presented him with a gift: they had captured a young gorilla alive.

It was between two and three years old, and it measured two feet six inches tall. They had found him in the forest, sitting on the ground eating berries. A few feet away sat his mother, also eating the fruit. The hunters shot the adult female, killing her instantly. "The young one, hearing the noise of the guns, ran to his mother and clung to her, hiding his face, and embracing her body," Paul later explained. When the hunters approached, the young gorilla let go of her body and raced up a tree; when they cut the tree down, they threw a cloth over the animal's head to capture it. Even so, the gorilla bit two of the men, and they struggled to keep hold of him. "He constantly rushed at them," Paul wrote. "So they were obliged to get a forked stick, in which his neck was inserted in such a way that he could not escape, and yet could be kept at a safe distance. In this uncomfortable way, he was brought into the village."

The sad story of the animal's capture didn't faze Paul, who was thrilled to have his hands on a living specimen. He described the moment as one of the happiest of his life. "All the hardships I had endured in Africa were rewarded in that moment," he wrote.

Paul named the gorilla Joe. Within two hours he and his men built a bamboo cage to house little Joe. He examined the animal and took detailed notes of everything he saw: the jet-black tone of his face and hands; the reddish-brown tint of the hair on the head; the short, coarse hair that covered the upper lip; the faint eyebrows, three-quarters of an inch long; the whiteness of the hair around his anus; the hair that hung from the wrist to the second joints of the fingers; the way the grayish hair of the legs grew darker nearer the ankles.

With Joe unhappily caged, Paul was seized by the desire to tame him. He began talking to him in a friendly way, getting close to the slats, but Joe rushed him and grabbed his pant legs, tearing them as Paul tried to move away. "He sat in his corner looking wickedly out of his gray eyes, and I never saw a more morose or ill-tempered face than had this little beast," he wrote.

Joe's attitude only worsened the next day. If Paul tried to come close to the sides of the cage, Joe threatened to attack. On the fourth day of his captivity, Joe pried apart two of the bamboo slats of his cage and escaped. Several men gave chase, while Paul ran toward his house to grab a gun. But when he ran through the open door, he was surprised to see that Joe had run into his house in search of refuge. Paul quickly shut the windows while others guarded the door. Eventually, they captured the frantic young gorilla with a large net and returned him to a newly reinforced cage. Joe was madder than ever.

Paul's attempts to train the animal were going nowhere. Joe would eat only fruits and leaves found in his native forest, which were difficult to collect. Paul tried to get him used to "civilized" foods, but the gorilla wouldn't touch them. For two weeks Joe sullenly resisted Paul's attempts to make peace. He gnawed another hole in his cage, attempted escape, and was again recaptured. Paul, deciding the cage was a lost cause, tethered him to a chain. He seemed to be doing better under this arrangement, and he began eating more of his preferred wild foods. But after ten days wearing the chain, he suddenly grew ill. Joe died two days later.

Paul captured another live gorilla—this one even

younger—a couple of months later. During a hike, he and his men found a baby male gorilla sucking at his mother's breast. One of the men fired. The adult gorilla slumped to the dirt. The baby clung to her, mouthing desperate cries as if trying to revive the mother's attention. Paul walked toward him, and the infant hid its head in his dead mother's chest.

Paul lifted the baby into his arms, while the men mounted the mother on a pole to carry back to their camp. When they arrived, they laid the adult on the ground, and Paul set the baby down nearby. "As soon as he saw his mother he crawled to her and threw himself on her breast," Paul wrote. "He did not find his accustomed nourishment, and I saw that he perceived something was the matter with the old one. He crawled over her body, smelt at it, and gave utterance from time to time, to a plaintive cry, 'hoo, hoo, hoo,' which touched my heart."

Without milk, the baby died on the third day. Paul preserved the body in a cask of rum.

THE MORE time Paul spent around gorillas, the more an uncomfortable notion gnawed at him.

These beasts seemed disconcertingly *familiar.*

When the gorillas reared up and ran on their two legs, they reminded him of men sprinting. Every time they roared, he thought he heard a human note somewhere inside that cry. Instead of a hunter, he at times felt like a murderer.

"Though there are sufficient points of diversity between this animal and man," he wrote, "I never kill one without having a sickening realization of the horrid human likeness of the beast." That likeness couldn't have been more grotesque—"a spirit of the damned," he thought.

By the time he departed Africa to return to America in 1859, Paul had collected more than twenty stuffed and preserved gorilla skins. Each of them possessed that eerie similarity to the human form, which is what gave them their unsettling power in the eyes of everyone who looked at them.

The disconcerting effect that a man gets when encountering his human reflection in nature had been noted twenty years earlier, when the uncommonly analytic young Darwin wrote about his voyage on the HMS *Beagle*. In South America, Darwin had been repulsed at the sight of a snake when he looked the creature in the eyes. Darwin wasn't content to let the discomfort he felt go unexamined. "I imagined this repulsive aspect originates from the features being placed in positions, with respect to each other, somewhat proportional to those of the human face; and thus we obtain a scale of hideousness," Darwin wrote in his book *The Voyage of the* Beagle.

On such a scale, Paul's gorillas were off-the-charts abominable.

"It was as though I had killed some monstrous creation, which yet had something of humanity in it," Paul wrote. "Well as I knew that was an error, I could not help the feeling."

He didn't know it, but at the exact same time that he was packing up his gorilla skins and preparing to leave Africa to return to New York, Darwin was putting the finishing touches on another book that would question whether Paul's feeling might not have been entirely wrong.

CHAPTER 16

Origins

London

Some modern writers have embraced the idea that Darwin's *On the Origin of Species* instantly became a publishing phenomenon in late 1859, devoured by average citizens in England and around the world. On the two hundredth anniversary of Darwin's birth, in 2009, magazines and newspapers recounted the "massive popular success" of a book that "sold out its first edition on the first day of its release." Although it's difficult to exaggerate the eventual impact of the book on the broad course of science and culture, it is easy to overstate the book's immediate, direct impact on the Victorian public.

The publisher John Murray originally printed about twelve hundred copies of *Origin*. But by the time the book's release date arrived, Murray had received wholesale orders from book dealers totaling about fifteen hundred copies. Today when people say that Darwin's book "sold out" on its first day of release, they're correct only in technical terms: it sold out to booksellers, not to readers. It took longer for those copies to get into the hands of the general public.

Most of the early readers belonged to a cloistered elite

of academics and gentleman scientists. Although this period marked the heyday of Charles Dickens and the rise of the popular novel, the era of general literacy hadn't fully matured by 1859. The average citizen of England wouldn't have read a book like Darwin's, and many would have struggled mightily to read any book at all. Almost half of British adults of marrying age couldn't even write their own names in the mid-nineteenth century. Darwin certainly fired up a small and hyper-educated pocket of the reading public in the months following his book's publication, but it took much longer for the blaze to spread among those who steered the course of conversation in pubs, on street corners, and in church pews.

IN THE book, Darwin didn't even address the notion of men evolving from apes, but it was implied. In *Origin*, he concluded that "probably all the organic beings which have ever lived on this earth have descended from some one primordial form, into which life was first breathed." Humans, in Darwin's view, were not exempt. According to his theory, man was not only related to apes but also related to nearly everything, from dogs to freshwater sponges.

Before he published the book, Darwin sent a copy to Richard Owen and shortly thereafter wrote him a letter. "I [should] be a dolt not to value your scientific opinion very highly," Darwin wrote. "If my views are in the main correct, whatever value they may possess in pushing on science will now depend very little on me, but on the verdict pronounced by men eminent in science. Believe me, yours very truly—C. Darwin."

Owen didn't like what he read. Although he accepted the idea of evolution up to a point, he was clearly uncomfortable with the broader implications of Darwin's theory. In an anonymous review of the book in the influential *Edinburgh Review,* Owen attacked Darwin's straight-line theory of evolution. Owen asked how, if all life-forms known to man sprang from the same matter at the same time, could simple organisms like protozoa continue to be found? Wouldn't they have "evolved" themselves out of existence? Owen also pointed out the absence of clear examples of transitional species, or "missing links," in the fossil record, a gap in the theory that Darwin himself had acknowledged.

Owen believed that instead of a single flash of creation out of which all life sprang, there had been numerous flashes. Those separate acts of creation might produce "archetypes" that followed their own processes of evolution to ultimately become what a divine Creator had ordained, Owen suggested. The single-celled organism that a modern scientist might view under a microscope represented a relatively recent "flash," Owen speculated, whereas the protozoan that eventually evolved into a man had begun its evolutionary journey much earlier, in the distant past. Owen also believed that sudden mutations—not Darwin's concept of very slow, gradual changes—might steer the development of a species. Unlike Darwin's theory, Owen's version of evolution was interpreted by many members of the clergy as preserving the biblically significant notion that humans were unique creations that developed independently of the other animals.

In the review, Owen anonymously stated that the book's valuable observations were "few indeed and far apart," and he said "most of Mr. Darwin's statements elude, by their vagueness or incompleteness, the test of Natural History facts." More than once in the review, he favorably referenced another naturalist who, unlike Darwin, wasn't so susceptible to wild conjecture—"Professor Owen."

———

ANOTHER ANONYMOUS review of *Origin* published in the London *Times* differed dramatically from Owen's.

"Mr. Darwin abhors mere speculation as nature abhors a vacuum," the review stated. "He is as greedy of cases and precedents as any constitutional lawyer, and all the principles he lays down are capable of being brought to the test of observation and experiment. The path he bids us follow professes to be, not a mere airy track, fabricated of ideal cobwebs, but a solid and broad bridge of facts."

The writer of this review was T. H. Huxley.

Like Owen, he'd been given a copy of the book by Darwin before it was published. On November 23, 1859—the day before Darwin's book went on sale—Huxley wrote Darwin a letter pledging to defend his theory, come what may. "I am prepared to go to the Stake if requisite," Huxley said.

Predicting that Darwin would be attacked by critics, Huxley vowed to swoop down on them with raptor-like ferocity. "I am sharpening my claws and beak in readiness," he wrote to Darwin.

When word spread among London's scientific community that Owen was the anonymous author of the *Edinburgh Review* article, Huxley targeted his prey.

·◦‖◦·

CHAPTER 17

In the City of Wonders

New York

The New York that Paul had returned to after he completed his African expeditions in late 1859 was a city pulled in two different directions. But at seven o'clock on the evening of December 19, thousands crowded the seats of the grandest opera house in Manhattan to try to nudge the city toward the one direction they favored: the South.

The rally at the Academy of Music, at the corner of Fourteenth Street and Irving Place, had been organized a week earlier by Democratic politicians with the backing of some of the city's most powerful business leaders. It was billed as a direct response to "outrages as those at Harpers Ferry" in Virginia, where the abolitionist John Brown had tried to start an armed slave revolt. Brown had been hanged days earlier, which put him on the fast track to martyrdom among those in the North who opposed slavery. But he was no hero to the twenty thousand people who had signed a petition backing this rally. The names on that list included about one-third of the city's eligible voters, and they represented a who's who of Gotham's mercantile elite.

Their support for the South was tied to money. In America's largest city, the seaport drove the local economy, and the cotton trade helped drive the seaport. The Southern states ferried their cotton harvest to New York, where it was shipped around the world. If the North and the South split, as more people were predicting in the wake of Harpers Ferry, New York might lose a key part of its business to rival ports, like Charleston and New Orleans. New York was vulnerable. Many who'd tied their fortunes to the cotton trade supported slavery under the assumption that the industry might collapse if slavery were abolished. If the industry collapsed, or if they were cut off from its harvests, they believed their profits might disappear.

When the opera house threw open its doors that night, about six thousand people jockeyed for seats inside the auditorium. Newspapers covering the event estimated that as many as fourteen thousand more huddled outside around bonfires, lighting fireworks and singing patriotic anthems.

"Hundreds and thousands are among us decrying the South and endangering our peace, but let us bless God that the bond of commerce binds us together," the former New York congressman James Brooks, one of the event's organizers, told the throngs. "Let us show the South that there are thousands, and hundreds of thousands, ready to stand by the Constitution and the laws."

One by one, a succession of orators took the stage and pandered to their fist-pumping audience, indulging its combative energy. Many earned easy applause by invoking the sanctity of the Constitution, which guaranteed protections for slaveholding Southerners. Several summoned religion to their cause, declaring that Abraham, Isaac, Jacob, and even Jesus Christ himself lived in slaveholding empires—and the

Bible never once explicitly condemned the practice. Charles O'Conor, a former U.S. attorney in New York, went so far as to suggest that slavery in temperate Southern states was a favor to blacks, and it showed consideration for "the condition the Negro is assigned by nature."

"Experience has shown that his class cannot prosper save in warm climates," O'Conor said of blacks, attracting roars of assent. "In a cold or even a moderately cold climate he soon perishes."

But in the streets of New York that evening, in the frigid December air, thousands of freed slaves were proving O'Conor wrong.

IN THE neighborhood of Five Points, about a mile and a half from the rally, fires from iron kettles made the streets shine and dance. Built on the swampy bed of a drained pond, it was America's most notorious slum, populated mostly by the poorest of recently arrived immigrants from Europe. Those unaccustomed to the stench that rose from its rotten foundations soaked handkerchiefs in camphor and pressed them to their faces before taking to the streets. Pickpockets ducked in and out of its saloons, dance halls, and gambling dens. Its narrow warrens earned nicknames that only thickened the atmosphere of sinister misery: "The Den of Thieves," "Murderers' Alley," "The Gates of Hell."

"Debauchery has made the very houses prematurely old," wrote Charles Dickens, when he visited the neighborhood (accompanied by a two-man police escort) in 1843. "See how the rotten beams are tumbling down, and how the patched and broken windows seem to scowl dimly, like eyes that have been hurt in drunken frays . . . hideous tenements which take their name from robbery and murder; all that is loathsome, drooping, and decayed is here."

A survey of the city's Sixth Ward, which was anchored by Five Points, showed that the neighborhood consisted of 3,435 Irish households, 416 Italian, 167 American-born, and 73 English. Many others were black, part of a population of freed slaves that numbered about twenty thousand in New York City. Few police roamed this district, so gangs—mostly Irish—with names like the Plug Uglies and the Dead Rabbits dispensed a homegrown version of rough justice. Abject poverty was an automatic assumption in Five Points.

When social critics of the day talked about amalgamation—the mixing of races—Five Points was often the place they turned to when arguing its downsides. In places like Almack's Dance Hall, the Irish jig and the African shuffle fused to create a new form: tap dance. Such unexpected combinations made Five Points one of the most vibrant, dynamic places in America in the early 1860s—the hottest recess in a boiling melting pot of cultures. But such hybrids were considered by many whites as little more than vulgar debasements.

William Bobo, a journalist from South Carolina, headed to Five Points to report on the dangers of mixing white and black cultures—then he decided his readers wouldn't be able to stomach his dispatch. "To give you a correct and critical description of the Five Points would only disgust, not benefit you," he wrote.

As early as the 1840s, Five Points became a magnet for white reformers. Protestant missions opened their doors here. Church leaders wanted to mop up vice, to cleanse the area of its sins. The most idealistic of them saw their work as a mission for the future of a city that was struggling mightily to determine what it was going to become.

One of these institutions, the Presbyterian Mission House, set up shop at 23 Centre Street, a nine-room house on the edge of the neighborhood. It was in those offices that John Leighton Wilson eventually came to work when he retired from his mission post in Gabon with a liver ailment. And when Paul Du Chaillu returned to America after his African adventure in late 1859, Wilson—who had recently moved back to his South Carolina home to attend to sick relatives—arranged for Paul to use the Mission House as his home base.

Here, in the middle of the most relentlessly bubbling melting pot in the world, at a time when everyone in America seemed to be debating the course of the future, Paul settled to forge his own.

Years later, an acquaintance recalled that people who saw him on the streets of lower Manhattan in late 1859 and 1860 were likely to mistake him "for a shoe clerk out of a job and on the verge of ejection from his boarding house." As an immigrant himself, in some ways Paul wasn't unlike many of the others he saw in Five Points. But unlike them, his mission contacts allowed him to step outside those dejected circles, to view the city from a privileged perspective. New York provided him with beguiling opportunities, and also vexing questions.

When introducing himself to people, should he emphasize his connections to America, where he sometimes told people he'd been born? Or should he claim affinity with France, or with Africa? With a storage box full of exotic animals that no one here had ever seen, should he cast himself as a scientist, or was it better to attract attention as a showman? In which direction did his future lie? How could he reconcile all the different elements that mixed together inside him?

For the next year, Paul searched for his own answers in

the ragged heart of a city whose future seemed just as uncertain as his own.

AT FIRST, he'd tried to play it straight. He hitched his future to science, to the learned societies that conferred respect on anyone associated with them. He'd gone to Africa with the support of John Cassin and Philadelphia's Academy of Natural Sciences, which was about as prestigious a body as could be found. But when he returned to the States, an uncomfortable realization began to dawn upon him: for some reason, certain members of the academy were acting as if they'd never heard of him.

They knew him well, of course. They'd been talking about his expedition during their meetings since 1855, and his name was scattered through the official transcripts of those gatherings. Cassin, who was Paul's point of contact, had first described him to the members of the academy as a young man who "possesses peculiar advantages as an explorer. He has lived long in the country, is entirely acclimated, speaks well two of the languages, and understands thoroughly the negro character. He proposes to proceed merely with convoys of natives from each tribe successively to the next." In other words, Paul's expedition had promised to be cheap, and the academy agreed to back him—a promise that Cassin conveyed to him in multiple letters. So Paul regularly stuffed and shipped animals from Gabon to Philadelphia. But the promised repayment never came. Suddenly the academy began ignoring his letters of inquiry.

In late December 1859, Paul wrote once again to remind the academy members that he'd returned to the States. He referenced additional letters he'd received from Cassin in

1857 that authorized him to continue with his explorations and indicated that part of his reimbursement would be sent immediately to John Leighton Wilson on Paul's behalf. But the money had never come.

Paul's letter was ignored. A few weeks later, he mailed another one. Then another. In January 1860, he sent them a detailed expense report along with excerpts of all the letters he'd received from the academy promising payment:

> *Under the circumstances of the case I shall be satisfied to receive the following sums:*
>
> *—Balance admitted to be due for birds taken by the Academy:*
> *$261.50*
> *—On account of expense of Camma expedition:* *$500*
> *—For birds lost 30 in number:* *$90*
> *—For boxes of land shells, 30 in number, 11 species:* *$15*
> *$866.50*
>
> *Mr. Cassin has informed me that since my return to this country he has sent 47 birds to Brenner (of which he gave me a catalogue) without consultation with me.*
>
> *I do not know exactly the terms upon which these birds were sent from the hall of the Academy and wish to know how the value of these birds may be secured to me.*
>
> *At the request of members of the Academy I have left a number of quadrupeds at the hall in respect to which the Academy will be pleased to let me know their determination.*
>
> *Very respectfully yours,*
> *P. B. Du Chaillu*

Around the same time that he was struggling to receive acknowledgment from Philadelphia, he was in contact with other prestigious academies. On January 5, 1860, he spoke to the American Geological and Statistical Society in New York. He was introduced to the group as a "French traveler who had advanced quite as far into the interior of the African continent as Dr. Livingstone." Paul displayed some props, including the skull of a gorilla, as he traced the route of his explorations on a large map. He explained that his strategy as a hunter was to wait until the chest-beating apes came close to him, then he opened fire.

"They died very easily," he said, earning a round of applause for what seemed like a display of admirable bravery.

He also traveled to Boston, to meet with Jeffries Wyman, the Harvard anatomist who'd named the gorilla and with whom Wilson had been acquainted. Paul gave him several gorilla skins and skeletons and the corpse of a juvenile gorilla soaked in alcohol, fit for dissection. Wyman in turn invited Paul to speak to members of the Boston Society of Natural History, where Wyman was president. Though his lecture charmed the members of the society—enough so that they later elected him an honorary member of the group—the event was sparsely attended.

As his efforts to penetrate the lofty realms of science and the academies stalled, Paul tried a new approach. Just two blocks west of the Mission House ran a colorful thoroughfare that had already become the prime destination for anyone hoping to earn a starring role in the American story.

He took his gorillas to Broadway.

AT THE beginning of 1860, the following advertisement ran in the pages of the *New-York Daily Tribune*:

DU CHAILLU'S AFRICAN COLLECTION, No.
635 Broadway, four doors below Bleecker St. Among which
is the gigantic GORILLA and a great many other specimens
of Natural History and Native Curiosities. Open day and
evening. Admission 25 cents.

The building at 635 Broadway had been constructed just five years before. It was marble fronted, with Italianate columns facing the street. Inside a long and narrow room, Paul unloaded all the cases he'd brought back from Africa.

Mounted fish. Birds wired with wings outspread. A chimpanzee skin that he labeled the "koola-kamba." A leopard. Spears and clubs he'd collected from the natives. The hollowed-out hippopotamus that he'd used as his "strong box" to carry dozens of his specimens home. And, of course, a stuffed gorilla—the most significant natural history specimen that had been unveiled in the country in decades.

"Hideous monsters with unearthly names beginning with four or five consonants, and dominions of outlandish places, such as the O-go-bai river, attract the attention," wrote a *New York Post* reporter after a visit to the newly opened showroom. "Uncouth forms of sluggish life repel the beholder. . . . But the guardian genius of the place is the gorilla, or man-monkey—one of the troglodyte tribe."

Paul's showroom seemed to have it all: an authentically newsworthy specimen of almost unimaginable novelty, a perfect location in the middle of the liveliest street in America, and a proprietor who wasn't afraid to take the floor and thrill anyone who'd listen with firsthand accounts of the beast's life in an impossibly romantic land.

All he needed now was an audience. But in a city torn apart by distractions, this would prove far from easy.

———

ONE PERSON who undoubtedly took notice of Paul's displays was P. T. Barnum. The world's premier showman had made a fortune introducing America to curiosities, both real and imaginary, and Barnum's American Museum on Broadway was a cathedral built to thrill. His ads during the first weeks of 1860 tempted the public with rare exhibits of natural history that included a marbled seal and two kangaroos. But now a newcomer had ventured onto Broadway with something that the press was calling a "man-monkey." Paul's gorilla threatened to steal Barnum's thunder, not to mention some of the quarters the public had to pay to get into either door.

Barnum was in a fix. He didn't have a prayer of acquiring anything that could equal the gorilla in terms of sheer novelty—that is, if he observed the rules of truth in advertising. But when it came to competing for the public's attention, Barnum rarely followed anyone's rules but his own.

Since John Brown's raid, Barnum had tried to capitalize on the racial tension that was making New York sizzle. In January he had begun displaying a wax statue of John Brown, an autographed letter from the abolitionist, and two spears that Barnum claimed had been used at Harpers Ferry. At the same time in his lecture room, Barnum staged performances of a play called *The Octoroon*, a tragic romance set in the rural South involving a woman whose father is a white plantation owner and whose mother is a mixed-race slave woman.

In the drama, written by an Irish-American Southerner named Dion Boucicault, the nephew of the plantation owner falls in love with the girl, whose name is Zoe. But her "dark, fatal mark" dooms them to heartbreak. In act 1,

Zoe explains: "Of the blood that feeds my heart, one drop in eight is black—bright red as the rest may be, that one drop poisons all the flood. . . . The one black drop gives me despair, for I'm an unclean thing—forbidden by the laws— I'm an Octoroon!"

The play struck nerves among audiences on both sides of the slavery debate. Abolitionists viewed it as a racist screed, while defenders of the South suspected that the play was thinly disguised abolitionist propaganda. "If such a being as Zoe ever existed in person, her mind and the taint of her blood would create a gulf between her and the whites that would be wider than the poles asunder," wrote New York's *Spirit of the Times* newspaper, which defended the South in the slavery debate. The play "is founded upon the false idea that there is an equality in the races, an idea that is preposterous, unnatural, and profane." Boucicault claimed impartiality: he was a Southern Democrat who felt "warmly towards the sunny South," he wrote in a letter printed in the New Orleans *Times-Picayune*, but hadn't intended his play to be mixed up in politics. But it was inevitable. Arguments defending and attacking the play ricocheted through the popular press, and Barnum reveled in the controversy.

But when Paul opened his exhibit down the street, Barnum knew he needed more incendiary firepower to attract crowds and blow his competition out of the water. In February, he took out an advertisement in the *Tribune* that ran down the entire length of the broadsheet page—an unusually large bit of self-promotion, even by Barnum's standards. At more than two thousand words, it was fifty times larger than Paul's ad in the same paper. What's more, many of Barnum's words were capitalized and attached to exclamation marks, drawing the eye to his side of the page:

That extraordinary living creature, just arrived from the wilds of Africa . . .

WHAT IS IT?

Is it a lower order of Man? Or
Is it a higher development of the Monkey? Or
Is it both in combination?

Nothing of the kind

HAS EVER BEEN SEEN BEFORE!

During the past six days not less than

25,000 PERSONS

have seen it, all of whom agree in pronouncing it the

MOST MARVELOUS CREATURE LIVING.

WHAT IS IT? WHAT IS IT?
WHAT IS IT? WHAT IS IT?

Barnum claimed that the mysterious creature, which could be viewed at the American Museum both day and night, had been "captured in the interior of Africa, on the borders of the River Gambia, by a party who were in search of the famous Gorilla."

The press in New York flocked to the exhibit, and Barnum manipulated the newspapers with masterly effect. Days after the exhibit's debut, Barnum was able to reprint the kind of press coverage most promoters could only dream about. All of New York's major papers published raves about the strange exhibit. The *Sunday Times* declared, "The new curiosity just added to the Museum stock seems to supply the real link between man and monkey." The *New York Times* added, "It seems to be playful as a kitten and imitative as a monkey. Singular facts about him—for he is obviously of the male species—are, that he can laugh with thorough heartiness, and occasionally mutter a few words of unintelligible gibberish. . . . It would be hard to find a place where more can be seen for a quarter of a dollar."

The truth, which the papers helpfully hid, was that the "nondescript" was a black man who suffered from microcephaly, a developmental disorder that results in an unusually small skull that tapers back at the forehead. Many microcephalics are severely mentally disabled, and their motor skills are often impaired, causing some to walk with difficulty. In a letter he wrote in April 1860, Barnum confessed that he purchased the man in Philadelphia from "a certain museum proprietor in St. Louis." Pictures taken of the man in early 1860 show that Barnum clothed his arms and legs in black fur (half of a monkey suit, essentially) and shaved his head, except for a small tuft on top, which heightened the sloping effect of his skull. When Barnum's manager described to the audience how the "creature" could walk upright with difficulty, the man would walk across the stage in what the *New-York Commercial Advertiser* described as an "exceedingly awkward" stride; when he'd stoop and rest his hands on his thighs, the announcer would tell the crowd that the gesture represented "the return of its desire to use all fours."

Barnum began to incorporate the "What Is It?" into performances of *The Octoroon*, trotting the man out onto the stage between the play's acts. Together, they formed Broadway's most popular attraction.

Paul and his very real gorilla, meanwhile, languished in comparative obscurity just a few blocks away, thoroughly upstaged by a living, breathing hoax.

··◦][◦··

CHAPTER 18

Fighting Words

Oxford, England

The famously gray skies of England opened up, and the leaders of Oxford University were thrilled that the sun was shining on the campus's latest addition: the Oxford Museum. The newly completed neo-Gothic structure was a cathedral to the natural sciences, a majestic assemblage of pillars, soaring arches, and spacious halls. The June 1860 meeting of the British Association for the Advancement of Science gave the university its first chance to show it off.

Lord John Wrottesley, the organization's president, welcomed attendees with a long-winded speech that threw obligatory nods to patrons and university dons and labored through a drowsy chronology of the most important British scientific advances during his own lifetime—an overview that wholly ignored Darwin's book. The omission might not have seemed especially blatant, because *On the Origin of Species*, though much on the minds of the scientific community, wasn't the book that was expected to stir up the most controversy among those in attendance. That distinction went to *Essays and Reviews*, an anthology of seven articles about the relationship between contemporary Christianity and science that had been published just three months before.

Essays and Reviews was written by liberal ministers within the Church of England, and it challenged the church to give science more freedom to delve into subjects that traditionalists considered sacrosanct. The essays urged a general retreat from literal interpretations of the Bible. Beliefs that did not square with the laws of nature—such as miracles—should be considered myths. In the twenty months after its publication, *Essays and Reviews* would sell more copies than *On the Origin of Species* would sell in its first twenty years. Two of its authors would lose their jobs in the church and be indicted for heresy.

Wrottesley, undoubtedly influenced by the interest that the essays had excited among clergy and scientists alike, assured the attendees of the conference that their work could never tarnish the glory of God.

"Let us ever apply ourselves seriously to the task," Wrottesley said, "feeling assured that the more we thus exercise, and by exercising improve our intellectual faculties, the more worthy shall we be, the better shall we be fitted to come nearer to our God."

Looking to the crowd, Wrottesley would likely have noticed Samuel Wilberforce, perhaps the most conspicuous attendee at the meeting. Wilberforce was the archbishop of Oxford, and he also indulged a personal interest in natural science. In the months before the conference he'd been meeting regularly with Richard Owen. With the anatomist's help, Wilberforce had even written an anonymous review of Darwin's book in the *Quarterly Review*. Owen, the man of science, was helping the man of the cloth develop arguments that each hoped would quash the notion that science could erode religion's authority. The meeting of the British association, which began in earnest the next morning, would put them both to the test.

THE MEETING'S first session that touched upon Darwin's theories ended with a debate about the gorilla. The subject was nowhere on the program.

It started innocently enough. The botanist Charles Daubeny presented a paper titled "On the Final Causes of the Sexuality in Plants, with Particular Reference to Mr. Darwin's Work." Darwin himself wasn't in attendance; he'd been suffering from what he termed "anxiety & consequent ill health" and had retreated from the public eye during those early weeks of summer. But Huxley and Owen sat in the audience, listening. As soon as Daubeny was finished reading, the man chairing the discussion asked Huxley, already established as Darwin's most energetic supporter, if he wanted to add more detail to defend the general crux of the theory. Huxley declined the offer. It wasn't the right forum for a debate, he explained, and Daubeny's paper hadn't really raised any ideas that needed defending.

Others jumped in to fill Huxley's silence. The discussion digressed into a jumble of loosely related subjects, from plant reproduction to primate behavior. Someone mentioned the gorilla. Owen, with his proprietary interest in the subject, rose to speak.

"Professor Owen wished to approach this subject in the spirit of the philosopher, and expressed his conviction that there were facts by which the public could come to some conclusion with regard to the probabilities of the truth of Mr. Darwin's theory," wrote a witness to the meeting in the next week's edition of the *Athenaeum*. "Whilst giving all praise to

Mr. Darwin for the courage with which he had put forth his theory, he felt it must be tested by facts."

Owen wanted to evaluate the theory by extending it to the question of man's origin. The test he had in mind was fairly simple: an anatomical comparison of humans and the recently discovered ape he believed represented man's closest relative. Owen identified the hippocampus minor, the spur today known as the calcar avis, as a defining difference—humans had it, he said, and gorillas didn't. The curvature of the hippocampus minor in the human brain, he said, was a key marker pointing to a vast difference in human and ape brain development. The sizes of brains were too dissimilar. The slow workings of natural selection, he suggested, could never turn a gorilla's brain into a human's.

Huxley couldn't hold his tongue any longer. He'd heard this argument from Owen before, and he remained unconvinced. He'd read papers by multiple anatomists that suggested that ape brains, particularly those of orangutans, *did* have a hippocampus minor, if only a rudiment of one. Huxley "denied altogether that the difference between the brain of the gorilla and man was so great as represented by Prof. Owen," reported the *Athenaeum*. Huxley "maintained that the difference between man and the highest monkey was not so great as between the highest and the lowest monkey."

The meeting broke up without incident shortly thereafter. But the polite disagreement would resume two days later when Wilberforce met Huxley.

FOR MOST of the twentieth century, history cast the verbal clash between Wilberforce and Huxley as a pivotal event—if not *the* pivotal event—of the early Darwinian debate. But as several modern historians have since pointed out, the most reliable evidence from the period, including all the eyewitness accounts and the letters written at the time by the participants, tells a different story.

The Saturday morning session began with another forgettable presentation, this one about the intellectual climate of Europe. Darwin was referenced. After the presentation, the chairman of the session called on Wilberforce to share his thoughts. The *Athenaeum* ran the most detailed account of his remarks, which argued that all the concrete facts available to science—from mummies found in Egypt to the sterility that afflicts hybrid species like the mule—speak to "the irresistible tendency of organized beings to assume an unalterable character." Darwin's book had presented a mere hypothesis, Wilberforce said, not a theory.

When the bishop was finished, Huxley rose to defend Darwin's ideas. He conceded that not every element of the theory had yet been confirmed, but he insisted that it was still the best theory concerning the origin of species that had ever been presented. Darwin's book was full of new facts and observations, all of which bore out the theory.

Although the *Athenaeum* and the other periodicals covering the conference didn't deem it important enough to mention, Wilberforce referenced Huxley's comments from two days before, when he had said that he believed men are more closely related to apes than apes are to lower monkeys. Wilberforce facetiously asked Huxley whether he believed he was descended from apes on his mother's side of the family or his father's. "This gave Huxley the opportunity of saying that he would sooner claim kindred with an Ape than with a man like the [bishop] who made so ill a use of his wonderful speaking powers to try and burke, by a display of authority, a free discussion on what was, or was not, a matter of truth," reported Alfred Newton, a zoologist who witnessed the exchange, in a letter written a month after the meeting.

In the mythology, Huxley's retort is a blunt smackdown, a blow that silenced the bishop with the force of incontrovertible truth. But descriptions of a decisive victory began appearing in recollections of the event written decades after the fact. The contemporary accounts presented the debate as anything but one-sided.

Huxley believed he had won the debate, but few others did. In a letter written just a week after the event, the director of the Kew Observatory said that Wilberforce emerged as the victor: "I think the Bishop had the best of it." Even Joseph Dalton Hooker, a good friend of Darwin and Huxley and a fellow defender of natural selection theory, believed Huxley's rebuttal, though justified in the wake of the bishop's "uglyness & emptyness & unfairness," was less than decisive. "Huxley answered admirably and turned the tables, but he could not throw his voice over so large an assembly nor command the audience." Wilberforce, in a letter written a week later, judged his debate with Huxley this way: "I think I thoroughly beat him."

From the distance of a century and a half, we can see that neither side registered a definitive, lasting victory during this debate. It would take decades before advances in the fossil record and new evidence from genetic theory helped another generation of scientists develop neo-Darwinism, the version of evolution that most scientists now accept.

Far from being the decisive word on the subject, the meeting only intensified the Darwinian debate. Both Huxley and Richard Owen worked harder than ever in the months following the Oxford meeting to prove their cases.

Owen would dedicate himself with unprecedented energy to what was called the gorilla question. In his mind, what he really needed to prove his argument was more specimens.

Within a couple of months of the meeting, he'd receive a letter that began, "My dear Sir, let me present you with a gorilla skin." The letter ended with the words "Sincerely, Paul du Chaillu."

⋅⋅◦][◦⋅⋅

CHAPTER 19

The Boulevard
of Broken Dreams

New York

While *The Octoroon* and the "What Is It?" exhibitions were at the height of their popularity, Abraham Lincoln checked in to the Astor House hotel, just across the street from Barnum's showplace. It was unseasonably warm for late February, and Lincoln, dressed in a new black suit, decided to go for a stroll with some young Republicans who were eager to show him the city.

Lincoln wasn't a recognizable face to New Yorkers in early 1860, and if he attracted any attention at all, it was much more likely to be a result of his unusual height rather than his celebrity. Lincoln and his small retinue of friends, three or four of them at most, on a Monday afternoon headed north for more than a mile. They passed Paul's museum, but it's not known if they stopped to peek inside. Lincoln's destination was just three doors up, at the corner of Broadway and Bleecker.

Lincoln strode into a photographic studio that was known as Broadway Valhalla. If a person wanted a photographic portrait taken of himself, he came here. Everyone from the current U.S. president, James Buchanan, to the "What Is It?" had stood in front of the cameras of the proprietor, Mathew Brady. Even Paul himself couldn't resist putting on his best three-button jacket and waistcoat and striking a pose in front of one of Brady's plush backdrops.

Brady's studio was, like Paul's showroom, a deep and narrow room designed to impress. The walls were covered in emerald wallpaper, and both the plush carpets and the buttoned leather divans were green and gold. A glass skylight overhead bathed the walls in natural light. Hundreds of portraits, mostly of famous faces and some colored with oils or crayon, hung in gilded frames. In one corner, Lincoln could have found a recent portrait of Stephen A. Douglas, the man whom he had run against the previous year in Illinois for a seat in the U.S. Senate. Lincoln's Republican Party in Illinois had won more popular votes during that campaign, but the Democrats won more seats in the legislature, which gave them the right to choose Douglas as the state's senator.

That race had thrust Lincoln into the national spotlight, and Republicans had begun talking about him as a potential candidate for the November election. But Lincoln remained little known in New York, a city that could make or break political fortunes. Almost 2.5 percent of the U.S. population lived in the city—more than lived in all but twelve of the country's thirty-four states. If Lincoln wanted to make an impression on the electorate, he needed to get his face out in the public. In 1860, that meant he needed a Mathew Brady portrait.

Brady welcomed Lincoln into his studio and quickly began sizing up the clean-shaven fifty-one-year-old lawyer with anthropological precision. He backed the camera away from Lincoln to turn the emphasis away from the facial details and toward Lincoln's towering stature. To make his narrow chest appear broader, Brady instructed him to open his long black coat, exposing his dark vest and the white V of his shirt. He arranged props beside him, including a fake pillar and a table piled with books, where Lincoln rested his left hand.

As Brady continued to examine him, he saw that something still wasn't quite right. The photographer suggested that Lincoln pull up his collar.

"Ah," Lincoln said, "I see you want to shorten my neck."

"That's just it," Brady said.

Lincoln stared into the camera with penetrating eyes and a clamped jaw. The image that Brady conjured suggested strength, learning, and purpose.

Later that same evening, Lincoln delivered a speech at Cooper Union in which he directly challenged the views that had been aired at the Academy of Music. The next morning, the newspapers devoted hundreds of favorable column inches to the event. "No man ever before made such an impression on his first appeal to a New-York audience," the *Tribune* concluded.

Brady's photograph was quickly reproduced all over the country, and it became the basis for a Currier & Ives print seen by millions. Within months, as tensions between the North and the South increased, Lincoln's name and image had spread enough to earn him the Republican nomination for president. "Brady and the Cooper Union made me president," Lincoln reportedly quipped later.

Paul might have been heartened by the rise of a man who penetrated the highest realms of power despite humble origins. Lincoln's authorized campaign biography of 1860 reported that his ancestry was shrouded in "incertitude, and

absolute darkness." To someone like Paul, who was growing adept at brushing aside inquiries about his past, Lincoln perfectly represented the romantic ideal of an American
meritocracy. After the 1860 campaign, Paul, whose interest
in politics was never strong, would claim until the end of his
life to be a proud and loyal Republican. But as Lincoln grew
into an iconic testament to the promise of self-determination,
Paul's experiences would quietly demonstrate how a deliberately shrouded past could come back to haunt a person in
unexpected ways.

ON THE rare occasions when the press took note of Paul's
collection on Broadway, the reports were at times indifferent
or dismissive. The *New York Post* drew the conclusion that
instead of inspiring wonder in the mysteries of far-off lands,
Paul's adventures underscored the benefits of a sheltered pro-
vincialism. "After viewing these monsters," the newspaper
reported, "the idea of a residence in the African continent is
received with coolness, and the beholder joins with fervor in
the exclamation of the persecuted Caddy [Jellyby], as, when
goaded to frenzy by the duties of the [Borrioboola]-Gha mission, she cried aloud, 'I hate Africa!'"

At the same time, his continued efforts to be reimbursed
by the Philadelphia academy slammed into an impenetrable
wall. In February, the academy finally referred the matter to
its curators. According to the academy's internal documents,
the curators decided early in 1860 that Paul's claim "in no
way concerns the Academy"—a judgment they didn't bother
to tell Paul. After he continued to inquire about the status
of his reimbursement, some of the members of the academy
asked the curators to outline the facts upon which they based

their conclusion, perhaps in an effort to put the matter to rest once and for all.

Enigmatically, the curators refused to give a concrete reason, saying instead that it was "inexpedient to report the facts."

In its letter to Paul, the academy didn't mention the curators' mysterious response. Instead, the members of the institution created a dubious technicality to disentangle themselves from any obligation to Paul. The letter explained that all of Paul's agreements had been with individual members of the academy—not with the institution as a whole. Therefore, the academy owed him nothing:

> A careful examination of its records shows that the Academy of Natural Sciences of Philadelphia has never engaged Mr. Du Chaillu to explore the Camma or any other country; nor has it contracted with him to furnish specimens of Natural History; nor has it ever authorized, by enactment of resolution or otherwise, any of its officers or members to make any engagement or contract of any kind or description with Mr. Du Chaillu. . . . It is regretted that Mr. Du Chaillu should have erroneously supposed that he was making explorations and collections in Western Africa under the patronage and at the cost of the Academy.

The rejection might have seemed an out-of-the-blue injustice to Paul, but whispers and rumors lay behind it. Could he have suspected that the "inexpedient" facts had something to do with his background, the details of which he had hidden so well since the day he had adopted the Wilsons as his new parents? If such suspicions had occurred to Paul, voicing them would only have drawn attention to that past. He dropped the matter, surrendering his claim to the $866.50—more than triple the annual wage of the average factory worker of the time.

Stripped of the money that he'd been counting on, and with only a struggling showroom to support him, he turned to his journals as a possible income source. He had met a physician and naturalist named Samuel Kneeland when visiting the Boston Society of Natural History; Kneeland, who had extensive experience in writing for scientific journals, had already helped Paul tame his imperfect English when preparing his lecture. Now Kneeland helped him transform the extensive journals from his expedition into a lively travel narrative. Paul worked feverishly, compiling nearly five hundred manuscript pages by the end of 1860. Harper and Brothers, a publishing company in New York, agreed to publish his narrative as a book.

But before that happened, Paul received an invitation from Richard Owen in London that would change his life. Wyman, the Harvard anatomist, had earlier written to his English counterpart encouraging Owen to meet with the young man and look at his specimens.

After being largely ignored for months by a distracted America, Paul—and his gorillas—got a second chance in a country that was free to give him its undivided attention and where he could again start anew.

--=] [=--

Part Two

CHAPTER 20

The Inner Circle

The London Richard Owen welcomed Paul into in early 1861 exuded magisterial confidence. Its residents had an iron certitude that *this*—a thirty-square-mile patch of soupy fog, hissing gaslights, and 2,803,921 human souls—represented the apex of civilization. Anyone new to London who might harbor the slightest doubt about the city's perceived preeminence needed only consult the very first paragraph of *The Popular Guide to London and Its Suburbs*:

> *London is the political, moral, physical, intellectual, artistic, literary, commercial, and social centre of the world. . . . Railways converge to it, and science, art, discovery, and invention seek it as their true home. Its merchants are princes, and the resolves of its financiers make and unmake empires, and influence the destinies of nations.*

A young man wishing to make a mark in the world could do no better than this. But just because it was a dominant city didn't mean that its power was conferred on just anyone, as London's smudged underclass—the ragpickers, costermongers, night-soil men, mud larks, shoeblacks, lamplighters, thimbleriggers—silently testified to every day. To tap into any of the city's power, a newcomer needed connections, someone to guide him over and around the invisible barriers of class and social rank. Someone who'd made that journey up the ladder of success himself. A person with friends in high places and a motive for sharing them.

Someone exactly like Richard Owen.

He was ruling class all the way. But even if John Edward Gray and his other rivals saw him as a pampered child of privilege, Owen considered himself a self-made man. He'd been born in Lancaster, the son of a merchant who occupied a social position a couple of rungs below the country's aristocratic elite. As a child, Owen discovered an interest in anatomy, and he collected the skulls of anything he could find—dogs, cats, mice, deer. As a teenager, he apprenticed with a surgeon and a pharmacist, which earned him the duty of tending to prisoners, both living and recently deceased, in the city jail and hospital. Within six months, a copy of a treatise called *Varieties of the Human Race*, which explored anatomical differences between the races, fell into his hands; on the same day, a black man died in the hospital. Armed with a paper bag and burdened by no ethical qualms whatsoever, Owen was determined to collect the skull, eager to measure its facial angles, the distance between ear and teeth, the brightness of the bone tissue—all details that scientists of the time believed distinguished blacks from whites. He snuck into the hospital's morgue, snatched the skull, and ran away with it.

Owen later studied medicine in Edinburgh before distinguishing himself as a gifted dissector at the Royal College of Surgeons. There he was given the responsibility of cataloging the college's museum collection—undissected specimens floating in pickling alcohol that had been collected by

England's most celebrated explorers, including Joseph Banks and Captain Cook. The college's most eminent surgeon, John Abernethy, took Owen under his wing and helped him climb to the top of Britain's medical establishment—the kind of act of generosity that Owen's mother had advised him to take advantage of, if it ever came his way. In a letter to her son, she said that great men shared a secret that lay at the foundations of their success. "Let them, if within their means and power, *become pupils of some person already eminent and in high repute*," she explained to her son. "By such a preparatory course they obtain two great objects—a well-grounded professional knowledge, and the opportunity of becoming known to all the friends and connections of their instructor."

It had worked for Owen. Through Abernethy, he became personal friends with celebrated scientists, including Cuvier, the world's foremost anatomist. One high-placed friend led to another. He worked hard, and his gifts for anatomical science impressed scientists throughout Europe. Eventually, he became the highest placed of them all.

When Paul showed up in Owen's life, the two men fell into instant companionship thanks to Paul's twenty-one gorilla specimens, which were the cargo of Owen's dreams. They formed an odd couple: one in his mid-fifties, the other in his twenties; one tall and professorial, the other a compact bundle of nervous energy; one thoroughly entrenched in England's ruling class, the other a total stranger to high society. It didn't matter that the young man had never set foot in England before. He had a friend.

Owen was a little like Wilson all over again: Paul played the son, Owen the father. Owen had clear motives for encouraging the relationship, but he could reassure himself that his attentions to Paul weren't wholly selfish. By promoting the

young man's career, Owen was repaying the favor that Abernethy had done for him when he'd sought a foothold in aca-
demia: "the opportunity of becoming known to all the friends
and connections of their instructor."

The young gorilla hunter quickly became a fixture at
Owen's home, Sheen Lodge. The people Paul met there held
the keys not only to the innermost chambers of Britain's scientific establishment. They held the keys to Victorian culture
itself.

RODERICK MURCHISON was fresh off his most successful year ever at the Royal Geographical Society. The organization had added 233 fellows to its ranks in 1860—the greatest
annual growth in its history. Its monthly meetings were more
popular than ever. When Owen made the acquaintance of a
young man who claimed to have traveled to parts of Africa
where no white man had ever set foot, and who'd seen things
none had ever lived to describe, Murchison jumped at the
chance to welcome him into the fold.

Murchison saw an opportunity to do what he'd done with
David Livingstone: create a celebrity who could get London
talking—about the Royal Geographical Society. Within
weeks of Paul's arrival, Murchison had cleared the decks at the
RGS's monthly meeting, announcing that the next featured
speaker would be an unknown adventurer who'd recently
explored Africa, and he'd be bringing his gorillas with him
to the meeting.

In the days leading up to the event, a newly elected RGS
fellow named Captain William Sandbach, who ran a West
Indies trading company, invited Paul to stay at his house,
located at 129 Mount Street in the posh Mayfair district.

Shortly after he moved in, a letter was delivered to the door. It was from the same publisher whom Murchison had contacted when he'd first encouraged Livingstone to write his book. The publisher, John Murray, had already been entrusted with a copy of the manuscript that Paul had sold to Harper and Brothers while in America.

Dear M. Du Chaillu,

My first impressions of your Adventures *have been confirmed and strengthened by further acquaintance with it. It is a most interesting and valuable record of enterprise and discovery in geography and natural history and reflects the utmost credit upon you.*

I shall be proud and happy to be the publisher of it and will use every effort to promote its success and to extend your Fame.

The terms which I propose are those on which I have brought out so successfully and so satisfactorily to the authors of The Travels of Livingstone, McClintock and others, viz. I will take upon me the entire cost and risque of the publication including the illustrations. . . .

Whatever be the time fixed for publication I would recommend that the printing be commenced without delay—I shall be anxious to hear from you in reply to this and meanwhile I leave to return your printed sheets in the hope of suggesting a few more corrections.

I am, my dear Sir,
Your ally and faithful
John Murray

Murray, in addition to being a friend of Murchison's and Owen's, was the most powerful literary tastemaker in London, and the drawing room of his house at 50 Albemarle Street was considered a nucleus of nineteenth-century literature. When Murray's father had launched the family publishing business, Lord Byron convened a literary circle that would meet at the house in the afternoons for tea, a group dubbed the "four o'clock friends." In subsequent years the Murray house published the novels of Jane Austen and Herman Melville, as well as the signature work of specialized experts from Malthus to Darwin. It had also enjoyed commercial success in a new genre that Murray liked to call "travellers' tales," which included the expedition narratives of Sir John Franklin and the African adventures of Livingstone. When Murray promised to promote Paul and his book, it wasn't empty rhetoric. As Ármin Vámbéry, a nineteenth-century Hungarian writer and traveler, said, Murray's house was "the literary forum of the elite," and any dealing with him automatically "raised an author to the position of a gentleman."

Paul wasted no time in making an appointment to visit Murray's place. On the same day he received the letter, he sent Murray a reply, carefully scripted in his best handwriting:

DEAR SIR,

Yours of this day has been received. In reply I will simply say that I accept the conditions you propose in your letter being those given to the most favored authors.

I certainly prefer to have you as the publisher of my work than any other man for I know the high integrity of your name and feel I am in the hands of one who shall not try to wrong me.

I am glad to hear that my work has interested you and I will feel rewarded for all the hardships I have endured if by my feeble means our knowledge of a hitherto unexplored region

has been advanced either in the geography or Natural History
of the country. . . .

As for the time of the publication we will arrange when
I shall see you; one intended to publish my work in America
this spring.

I would like to see you and I will call at your office
tomorrow at 2 o'clock in the afternoon. In case you could not
be there, please write me a note.

Yours very truly,
P. B. Du Chaillu

Despite Paul's agreement with Harper and Brothers in the United States, Murray fast-tracked publication of the book to ensure that it would appear in England first.

Even before Paul had made his first public appearance in London with Murchison at the RGS, the talk began circulating among the city's chattering classes that a new adventurer was in town and that his story demanded attention.

CHAPTER 21

The Unveiling

The storms that had broken windows and scattered shingles had finally moved on, leaving behind a damp, windless chill. Slowly, the city's doors began to open, and those who'd spent days huddling inside their homes stepped out, testing the weather. Steam rose from the muzzles of the horses in the streets, but the cold was tolerable—nothing a heavy cape or a couple of petticoats couldn't handle. The gas lamps flared a pale yellow, throwing long shadows across the wet cobblestones. By early evening, a cast of thousands was drawn mothlike into the city's social swirl.

Bonneted women in crowded omnibuses bumped shoulders in rhythm to the hollow clop of horse hooves. Aproned servants steered handcarts through the traffic. Men stumbled out of the smoky private clubs on St. James's Street, joining a pedestrian flow that seemed headed in only one direction: toward Piccadilly Circus, the teeming roundabout that united Regent Street and Piccadilly.

This was the beating heart of London's West End, and the neighborhood offered something for anyone lucky enough to have a few shillings rattling around in his pockets on this Monday evening, February 25, 1861. Two blocks north on Drury

Lane, Charles Kean, one of the most celebrated stage actors alive and a personal favorite of Queen Victoria's, was getting ready to take the stage as Hamlet. A block to the west, Belgium's Henri Vieuxtemps, the greatest violinist in the world, was about to perform Mozart in B-flat. At the Egyptian Hall, the actress and singer Emma Stanley was preparing to unveil a one-woman musical extravaganza that required her to make no fewer than thirty-seven complete costume changes.

But those were sideshows, really. The main event—the hottest ticket in town—was about to start inside Burlington House, the enormous Palladian-style mansion that dominated Piccadilly.

The private carriages that stopped in front of the building's imposing brick gate unloaded a very particular breed of Londoner on this evening. Their clothes were designed not to merely fend off the cold but to be *seen*. The women wore evening dresses, puffy confections of crinoline and silk. The men wore black dress coats with velvet collars and fashionably wide lapels. Many had traveled in straight lines down Piccadilly from the stone mansions that lined Green Park. These were London's aristocrats: lords and ladies, captains of industry, international statesmen, and intellectuals poised on the leading edge of science.

But tonight they were content to be mere spectators, to fade into a crowd, and to be wonder-struck.

INSIDE BURLINGTON House's west wing, Paul clacked across the marble floors toward a stately grand hall. From high oak-paneled walls, enormous portraits of the men who put England at the symbolic center of the universe looked down with unblinking severity. Lion-maned heroes with penetrating eyes. Christopher Wren, Isaac Newton, Joseph Banks—giants who'd lifted English culture to unprecedented heights through the force of intellect and discovery. Each of them had, at various times, presided over one of the learned societies that anchored London's intellectual life. Several of those institutions—the Royal Society, the Linnean Society, and the Royal Geographical Society—were now based here, inside this mansion. The men looking down from the walls were legends, and Burlington House was their legacy.

Everything about the building, from the fluted Corinthian columns to the ornate chandeliers, exuded a regal confidence, and everyone pouring into the mansion's great hall seemed right at home within that atmosphere.

Everyone, that is, except Paul.

He was embarrassingly unfamiliar with the rigors of Victorian etiquette. His English was tainted not only with French inflections but also with liberal doses of American slang. He flitted among the bystanders looking very much like the teenager washed up from the river. Even at the age of twenty-nine, he stood about five feet three and weighed just over a hundred pounds. Surrounded by the wintry complexions of Britain's elite, his face looked like that of a sunbaked kid who'd spent too much time playing on a beach. His dark eyes were forever in motion, drinking in details, twinkling with something acquaintances invariably labeled "child-like wonder." Almost everywhere he went in London—but never more so than here in Burlington House—he seemed a boy among men.

From a small riser at the front of the hall, he looked out over the multitudes. The room was built to hold up to a thousand people, and it was packed to capacity. The crowd was thick with "savants"—those Victorian luminaries who were responsible for a frenzied percolation of ideas that would challenge the fundamental assumptions upon which Western society had been built. They were in the process of becoming legends, destined to take their places among the immortals on the walls. Together they were creating a new language for the whole world. Owen, who'd coined the term "dinosaur," sat near the front of the room. Huxley, credited with inventing the word "agnostic," lingered within striking distance. So did Francis Galton, who came up with the concept of eugenics. William Gladstone, who would lead England as prime minister four times, remarked that evening that he felt like "the lowest schoolboy in the school" when gazing upon the assembly.

Paul struggled to control his nerves. His name was called. He walked to the center of the rostrum.

HE BEGAN to speak but faltered. He sheepishly admitted to the audience that he felt a bit overwhelmed. He said that as a New Yorker, he couldn't help but feel out of place. He knew it was an impossibly distinguished crowd, but the only person he recognized by sight was George Mifflin Dallas, who'd been the American vice president under Polk and was now the ambassador in London. "I trust he will offer me his protection," Paul told the audience, risking a little humor.

His humble admission of discomfort charmed the audience. As he continued to speak, some of his inhibitions began to melt. Animated gestures gave shape to his words. He slipped into the narrative flow of a natural storyteller.

PAUL'S IRREPRESSIBLE energy, even when it wasn't fueled by anxiety, often gave him the appearance of someone who was seconds away from boiling over with excitement. His hands rarely kept still, and his eyes never did. Young people loved him, seeing something playful inside him that invited connection. Owen's children swarmed him when he visited Sheen Lodge, and Murray's kids would beg the "Monkey Man" to tell them stories of the jungle. He always obliged, whispering when suspense required it, exploding in a pantomime of ferocity when their anticipation reached a breathless peak. One of his friends in London, Edward Clodd, remembered how Paul could hold his children spellbound with his animated tale of encountering the gorilla in the wild for the first time. "My children will never forget his telling it to them," Clodd recalled. "His vivid imitation of the awful roar of the animal as he beat his breast with his huge fists, and of the terrible human groan with which he fell prone on his face, made them shriek with fear, so realistic was it." Unlike most adults, Paul seemed to lack the veils that separated inner emotion from outward expression. His mix of vulnerability and indefatigable optimism could charm even the hardest hearts. Anyone with a natural tendency to root for the underdog tended to feel some indefinable tug toward him.

At Burlington House, he unloosed that infectious charm on the crowd. In his quirky accent, he described the dangers he faced with an élan that flirted with false naïveté: it was as if encounters with unknown beasts and deadly serpents were inevitable hurdles in the life he had chosen and therefore didn't warrant complaint. His youthful appearance and diminutive size made his hair-raising tales seem all the more dramatic. He wasn't above mocking himself, and the audience rewarded him with laughter and applause. He was, in one magazine's words, "almost the last man whom one at first sight would set down as a great explorer, adventurous traveler, and a naturalist of no ordinary attainments."

During his first lecture in England, his stage props boldly underscored the startling disconnect between the stoutness of his adventures and the frailness of his frame.

Workmen had lugged the props into the building, two bulky parcels swaddled in coarse drugget. They eventually unwrapped them on the riser behind Paul like fragile mummies recovered from dusty tombs. Under those layers of rough fabric, the parcels were revealed as oddly misshapen assemblages of arms, legs, teeth, and fur.

They were stuffed gorillas, two full-grown adults positioned in attitudes of diabolical menace. No one in the crowd had ever seen anything quite like them, and they didn't simply attract attention; they commanded it. Compared with the impish little man who lectured in front of them, they appeared to prove the idea that the monsters of nightmare really do exist.

THE AUDIENCE was hooked on Paul's every word as he described how he encountered the beasts and slew them. It didn't matter that the gorillas had been displayed before in the United States. The lecture at Burlington House was the animal's true unveiling—its red-carpet debut.

As soon as the applause for Paul died down, Richard Owen made his way to the podium to make sure that no one in the room had missed the significance of what they'd just witnessed—"the most strange and extraordinary animal of the brute creation."

Before that night, the gorilla had been largely a mystery whose nature could only be guessed at. But now, Owen said, this young man had opened the world's eyes to a wonder of the natural world that could not be more relevant or timely.

"In natural history, as we go on comparing form with form, of course we soon become impressed with the idea of a connected scale, and the interest increases as we ascend," Owen explained, dropping a reference to the theory of evolution. "But when we come so near to ourselves as we do in the comparison of this tailless anthropoid ape, the interest becomes truly exciting."

The crowd continued to stare at the gorillas, but if they looked very closely, they might have noticed that something was missing, and it spoke to Paul's uncertainty addressing such a terrifyingly *respectable* audience. There were dozens of ladies in the lecture hall—unusual for the era, but indicative of the interest that Murchison and Owen had whipped up— and their attendance had concerned the young speaker.

Fearful of offending, he had modified his exhibits to conform to what he assumed were proper codes of display in Victorian England.

He had castrated them.

THE NEXT day, when Paul and his friends looked to see what the press had written about the meeting, they discovered that the event had been almost completely overlooked. The *Times*, the paper of record in London, gave it only the slightest mention two days later.

A few days after that, an anonymous reader identifying himself as "one of the oldest Fellows of the Royal Geographical Society" published a long letter to the *Times* to remedy the oversight. He wrote that he'd never attended a meeting at which more interest had been excited and that this "citizen of New York" deserved recognition on both sides of the Atlantic.

Well-intentioned though it was, the letter proved unnecessary. Others were making more direct appeals for the newspaper's attentions. The evening after Paul's lecture, the editor of the *Times*, John Thadeus Delane, received a dinner invitation from Roderick Murchison.

When Delane arrived at Murchison's home in Mayfair, he found a dinner table crowded with luminaries. The Duke of Wellington was there. So was Owen. Joseph Dalton Hooker, who happened to be Charles Darwin's best friend in addition to being a preeminent botanist, had also arrived. The guest of honor, the one Murchison wanted them to meet, was Paul.

Delane was famous for keeping his cards close to his vest, rarely revealing what he thought of an issue or a person when in public, content to let his newspaper speak for him. The newspaper's voice was his voice, and it was one of incontrovertible authority. Anthony Trollope created a recurring character based on Delane in his Barsetshire novels: Tom Towers, the editor of the *Jupiter* newspaper. According to Trollope, Delane sat in his office like a god atop Mount Olympus, "where, with amazing chemistry, Tom Towers compounded thunderbolts for the destruction of all that is evil, and for the furtherance of all that is good, in this and other hemispheres." His opinion, so difficult to predict and so warily anticipated, was the only one that truly mattered throughout the British Empire. Trollope wrote that politicians yearned for his approval, the church feared him, and generals based their strategies less on their enemies than on what the editor might write about them. "It is probable," Trollope wrote, "that Tom Towers considered himself the most powerful man in Europe; and so he walked on from day to day, studiously striving to look a man, but knowing within his breast that he was a god."

After that dinner, Delane didn't reveal what he thought of Paul. But with Murray rushing to publish the adventurer's book, Delane's newspaper would get a chance to publicly judge the unknown young man who'd dined with so many bold-lettered names.

Within weeks, Paul would become a bigger celebrity than any of the other men who'd been sitting around that dinner table.

⋅⋅∘]|[∘⋅⋅

CHAPTER 22

The Great White Hunter

On a spring morning in Bloomsbury, deliverymen began hauling several heavy boxes into Mudie's Lending Library.

Huffing their way up the iron staircase, they could look down upon the library's grand salon and get a bird's-eye view of one of the liveliest businesses in the city. Men and women shouldered past one another through the door that led out to New Oxford Street. Dozens scoured the bookshelves for new titles, while others clamored around the front desk. It was a typical Saturday morning in the salon. Which is to say it was a madhouse.

Charles Mudie, an aptly bookish man of forty-two years, lorded over this domain with the understated authority of someone with nothing to prove. His father had made a living selling magazines and secondhand books, and Mudie had been expected to follow in his footsteps as a traditional newsagent. But by the time he was twenty-three, he had completely transformed the family business—and the entire publishing world—by creating a lending library that would dictate the reading habits of Victorian masses for decades to come. If you asked someone in London to name the most influential man in the publishing world, you'd likely get one of two answers: those who knew anything about literature would say John Murray; those who knew anything about business would say Charles Mudie.

Mudie's big idea had been deceptively simple and absolutely irresistible for those who found the prices of books prohibitive: for a subscription fee of just a guinea—a little more than one English pound—a person could borrow an unlimited number of books during the course of a year, provided he or she checked out only one at a time. It was a business model that required a high volume of stock and steady flow of customers. Mudie had been wildly successful in securing both. In 1860, with business booming, he was able to open new branches in Manchester, York, Birmingham, and several other English cities.

At his flagship location on New Oxford Street, the Ionic columns that stood in the main salon were mere decoration; what really supported the place was the *books*, which rose to the ceiling against every wall and numbered no fewer than 800,000 volumes. Light iron walkways allowed customers to browse the higher shelves. The long counter near the door was swarmed by subscribers returning books or checking out new titles. An average of 3,000 volumes changed hands over that counter each day. The books that had suffered too much wear and tear were tossed into a pile bound for the on-site "infirmary," where broken spines were mended and ripped covers replaced. Those beyond hope were exiled to the "charnel house," to be ground to pulp and mixed into manure.

This was where almost everyone in Victorian London, with the notable exception of the wealthy, got their books. The market dominance of this single library meant that publishers catered to Mudie's demands, and he exploited that power. Because he bought hundreds of copies of titles that he predicted would be popular, Mudie had the ability to single-handedly determine a book's print run.

When Mudie recognized an increasing public demand for long novels, he demanded that publishers break those stories into separate volumes, which enabled him to speed up his turnover and fulfill the demands of more of his subscribers. In this way, Mudie invented the "Victorian three-decker." He profoundly influenced the way many of the most influential novelists of all time structured, plotted, and paced their stories. He also forced publishers to charge the public artificially high prices—a scheme from which only Mudie was exempt. An average three-decker cost the reading public more than thirty-one shillings; Mudie paid just fifteen shillings.

The publishers played along because it was the only way their books might find a place on Mudie's advertised list of "Principal New and Choice Books in Circulation." It was the Victorian version of the best-seller list, and it was determined by both popular demand and Mudie's own sense of what was worth reading. If he considered a book salacious or improper, he banned it from the list.

Since Mudie opened his lending library in 1842, a handful of titles had distinguished themselves as noteworthy successes, capturing the fancy of both Mudie and his loyal subscribers. The first was Macaulay's *History of England*, which appeared in 1849. To fulfill demand, Mudie alone bought 2,500 copies of the book—more than most books sold in total. In 1858, Livingstone's *Missionary Travels and Researches in South Africa* superseded it, with Mudie eventually buying 3,250 copies from Murray and circulating them to an estimated thirty thousand readers.

On this Saturday morning, the boxes lugged upstairs to

Mudie's storage room were filled with *Explorations and Adventures in Equatorial Africa*. Mudie, who'd had good luck with travel narratives and took a personal interest in them, ordered an unusually large initial quantity from Murray: 500 copies. He immediately cleared space to display the title on a prominent shelf.

A long, jagged stripe of the new books ran halfway across the main salon. As more men and women reached for the copies, Mudie watched the length of the stripe steadily dwindle. On Monday, he sent a message to Murray informing him that he wanted to order 250 more copies. After that order was filled, Mudie on Wednesday requested another 500. On Friday, he ordered another 250 copies.

The book hadn't been out for a week, and every customer who poured into Mudie's seemed to have the same request: "Du Chaillu's gorilla book." It shot to the very top of Mudie's advertised list.

It wasn't just the library's most popular book of the season. It was fast becoming one of the most popular in Mudie's history.

AMONG THE first publications to review the book was John Thadeus Delane's *Times* of London. The article, embroidered with superlatives and running more than three thousand words, brought the mysterious Delane's view of Paul's adventures into clear focus.

"We must go back to the voyages of Le Perouse and Captain Cook, and almost to the days of wonder which followed the track of Columbus, for novelties of equal significance to the age of their discovery," the *Times* stated. "M. du Chaillu has struck into the very spine of Africa, and has lifted the veil of the torrid zone from its western rivers, swamps, and forests."

The review summarized his travels and quoted liberally from the text. It marveled at Paul's encounters with cannibals and especially with the "interesting monster" that currently had the public in its terrifying grip. The man who had journeyed so far to tell such incredible tales, the article suggested, had shown a rare sort of courage that deserved special celebration.

"Such exploits on the part of a slight, wiry American gentleman of French extraction, whose modesty and evident trustworthiness have commended him to almost everyone who has met him in English society," the newspaper concluded, "are a sufficient explanation of the eagerness to read his narrative, and more than an excuse for our own prompt attention to its varied contents."

Competing periodicals heaped similar praise upon the book. The *Saturday Review* concluded that "M. du Chaillu's narrative will not disappoint the expectations which it has excited," and the *Spectator* reported that the book was "all that could possibly be wished." Many of the reviews stretched on for thousands of words, excerpting Paul's passages and then lamenting the fact that they couldn't print more. The book reviewer for the *Critic* expressed his dilemma this way: "To quote everything that is interesting in this volume would be tantamount to a reprint of the entire volume."

EXPLORATIONS AND Adventures in Equatorial Africa as originally published in England ran to 479 pages, including seventy-three sketches and a map that roughly charted Paul's expedition routes. The frontispiece was an 8.5-by-11-inch foldout drawing of a gorilla standing upright with its right foot propped on a rock. The posture was deliberately human. The artist who'd drawn the portrait made sure that a conveniently placed tree branch performed the work of Adam's proverbial fig leaf, artfully obscuring the gorilla's groin.

The book was presented as a strictly chronological reconstruction of Paul's experiences in Africa between 1855 and 1859. Aside from the members of a handful of African tribes, Paul was the first person to ever encounter a gorilla in the wild, and he clearly understood that his book's success rested on his descriptions of that animal. He rarely wasted an opportunity to titillate. His gorilla was a "hellish dream creature" that rules the jungle through sheer intimidation.

As the title suggests, the book had competing aims: at times it read like an analytic document of *exploration*, and at others it switched narrative gears into a boyish *adventure* that tantalized with the thrill of the exotic. The book's schizophrenic desire to have it both ways threatened at times to tear the narrative apart, but that tension also lent it a propulsive energy that made it so popular.

Was Paul condemning the ferocious, terrifying natural world that he encountered as a place that needed to be civilized, or was he celebrating it? Did he believe the gorilla bore a striking resemblance to man, or did it represent the very opposite of humanity? He left the questions open, and the resulting uncertainty shook the ground under the feet of Victorian readers, providing them with the irresistible jolt of a thrill ride. The world had entered an era of uncertainty, and Paul's own conflicted vision supplied it with a fascinating, if disorienting, fun-house mirror.

He portrayed himself as a young man who was eager for

a challenge and fortunate to survive the experience. In the preface, he offered this summary of his journey:

> *I travelled—always on foot, and unaccompanied by other white men—about 8000 miles. I shot, stuffed, and brought home over 2000 birds, of which more than 60 are new species, and I killed upwards of 1000 quadrupeds, of which 200 were stuffed and brought home, with more than 80 skeletons. Not less than 20 of these quadrupeds are species hitherto unknown to science. I suffered fifty attacks of the African fever, taking, to cure myself, more than fourteen ounces of quinine. Of famine, long-continued exposures to the heavy tropical rains, and attacks of ferocious ants and venomous flies, it is not worth while to speak.*

The book dripped with sensational descriptions of gorillas, a beast whose chest beating could be heard a mile away and whose hideous roar made the trees tremble. Tellingly, the story of the gorilla bending the native hunter's gun barrel in his mouth was highlighted in all its monstrous glory, even though Paul made it clear he didn't witness the event himself. A pencil sketch of the imagined scene wrung every drop of melodrama out of the anecdote.

But near the end of the book, the lurid tone that defined most of the narrative abruptly gave way to a healthy dose of analytic sobriety. Despite the brute's reputation as a man hunter, Paul described the gorilla as shy and a "strict vegetarian." He noted, "I examined the stomachs of all which I was lucky enough to kill, and never found traces there of aught but berries, pineapple leaves, and other vegetable matter." He continued:

> *I am sorry to be the dispeller of such agreeable delusions;*
> *but the gorilla does not lurk in trees by the roadside, and*
> *drag up unsuspicious passers-by in its claws, and choke them*
> *to death in its vice-like paws; it does not attack the elephant*
> *and beat him to death with sticks; it does not carry off women*
> *from the native villages; it does not even build itself a house of*
> *leaves and twigs in the forest-trees and sit on the roof, as has*
> *been confidently reported of it. It is not gregarious even; and*
> *the numerous stories of its attacking in great numbers have*
> *not a grain of truth in them.*

Such abrupt shifts between sensationalistic hyperbole and dispassionate debunking epitomized the book's split personality. The schisms were most obvious when Paul struggled with a core question: Did the gorillas bear a strong relation to men, or did they not?

As might be expected of someone who owed his education to missionaries, he felt a strong pull toward religious tradition. He'd been taught to view humans and the animal kingdom as definitively separate, and he repeatedly professed a firm belief in that idea. In a passage that was added to the book just before its publication, he contrasted the cranial capacity of the skulls of men with those of apes. Paul asserted that the differences evident between the skulls provided "incontestable proof of the great ascendancy of the intellectual life of the human species, even in the lower orders of the human family. . . . The difference of size of brain or cranial capacity between the highest ape and the lowest man is much greater than between the highest ape and the lowest ape." The addition had Richard Owen's fingerprints all over it. It sounded as if it could have come straight from one of the anatomist's lectures.

Despite the book's respectful nods to convention, it was difficult to determine from the text whether Paul truly believed it. To him, the gorilla represented "a being of that hideous order, half-man, half-beast, which we find pictured by old artists in some representations of the infernal regions." He wrote:

> *I protest I felt almost like a murderer when I saw the gorillas this first time. As they ran—on their hind legs— they looked fearfully like hairy men; their heads down, their bodies inclined forward, their whole appearance like men running for their lives. Take with this their awful cry, which, fierce and animal as it is, has yet something human in its discordance, and you will cease to wonder that the natives have the wildest superstitions about these "wild men of the woods."*

Explorations and Adventures isn't just the story of Paul's quest "to find the very home of the beast I so much wished to shoot." It's also the chronicle of a conflicted man's attempt to suppress a terrifying suspicion: the beast he's pursuing seems, in a most unsettling way, to resemble himself.

SEVERAL WEEKS after his lecture at the Royal Geographical Society, Paul delivered a speech at the Royal Institution, where he again confronted a lecture hall that was "crowded to excess."

Owen sat next to his wife, Caroline, in the audience. She described the event in her diary: "M. Du Chaillu gave a very quaint, clear, and interesting account of his travels in Africa, and his meeting with the gorillas, a row of which hideous creatures was overhead: some skulls were before the lecturer, who traced his progress on a large map as the lecture proceeded."

As he continued to lecture throughout London, Paul would survey the crowd, gauge what sort of story they wanted to hear, and then he'd give it to them. Anyone who regularly attended such meetings had heard enough academic lectures to last a lifetime. But stories that thrilled, delivered by someone who had the authority of a man of science, were a relative novelty. His audience wanted to be entertained.

"Shooting a lion," he told the Royal Society audience with nonchalance, "is merely shooting a great beast. But there is something more dreadful in killing one of these apes. Their death is terrible, and, when in the conflict, you can but feel that if you make any mistake with the monster, he will not make any mistake with you."

He wasn't just walking the thin line between credibility and bravado; he was dancing on it. Whether or not he realized it, he was turning himself into a caricature—a fearless explorer who had faced nature's darkest secrets and lived to tell about it. The image was destined to outgrow the man himself. In more ways than one, Paul was creating a monster.

·◦❘❘◦·

CHAPTER 23

Into the Whirlwind

May in London marked the beginning of a period known simply as the Season. The gray skies rolled back to welcome the sun, and the city swelled with an influx of aristocracy who each spring moved out of their countryside estates and into their town houses in Mayfair and Green Park.

Ladies with parasols shopped at the arcades in the mornings, and they "paid calls"—short home visits of fifteen to thirty minutes—to friends between 3:00 and 6:00 p.m. The House of Lords was in session, and its members spent almost as much time engineering private business deals as they did crafting legislation. Private balls and soirees consumed the evenings. The theater season kicked into full gear.

To take advantage of the influx of social traffic, the Royal Geographical Society opened the doors of its headquarters at 15 Whitehall Place, inviting the public to visit an exhibition of one of Paul's gorillas. From 3:00 to 5:00 p.m., visitors could stand in front of the brute, whose arms were spread in a posture of fearsome attack. The animal's lips were pulled back to better reveal the teeth, which, according to one reporter, "from their size might fairly be denominated tusks." The label on the specimen read, simply: "KING."

Instead of returning to the United States, Paul opted to remain in the house at Mayfair belonging to Sandbach, the RGS fellow. Paul occasionally visited the exhibit himself, amazed at the adulation that was showered upon him by some of the wealthiest people in the world. He dined with lords and ladies, dukes and duchesses. His name seemed to be on everyone's lips.

If Paul himself wasn't the most popular attraction of the season, the title had to go to the animal he had brought to the world's attention. The distinction made little difference. The man and the beast had become inextricably linked in the popular imagination.

More than any animal before or since, the gorilla had become an instant cultural phenomenon, dominating every level of public discourse from the highest of the highbrow to the lowest of the lowbrow.

BEFORE THE 1840s, the word "cartoon" wasn't associated with humor or caricature. A cartoon was simply what artists called the preliminary sketches they used as guides for paintings, frescoes, or tapestries.

But in 1843, London's *Punch* magazine poked fun at Parliament by sketching a "cartoon" of a possible painting that could be hung in the Palace of Westminster for an upcoming state-sponsored art contest. After this, the word "cartoon" stuck as a label attached to humorous sketches, like the ones *Punch* began to publish each week.

In the spring of 1861, one of the magazine's cartoons pictured a gorilla standing upright, holding a walking stick, and wearing a placard around its neck that read, "Am I a Man and a Brother?" Everyone who read *Punch* would have understood the joke instantly. Before England had abolished slavery in 1833, the symbol of the Society for the Suppression of the Slave Trade was a portrait of a slave, chained and kneeling in a posture of supplication, bearing the inscription "Am I not a man and a brother?"

Punch titled the cartoon "Monkeyana" and featured a long poem as its caption. Signed by an author called only "Gorilla," the poem occupied three-fourths of a page:

Am I satyr or man?
Pray tell me who can,
And settle my place in the scale.
A man in ape's shape,
An anthropoid ape,
Or monkey deprived of his tail?

The Vestiges taught,
That all came from naught
By "development," so called, "progressive";
That insects and worms
Assume higher forms
By modification excessive.
Then Darwin set forth,
In a book of much worth,
The importance of "Nature's selection";
How the struggle for life
Is a laudable strife,
And results in "specific distinction."

Let pigeons and doves
Select their own loves,
And grant them a million of ages,

Then doubtless you'll find
They've altered their kind,
And changed into prophets and sages.

The poem continued for several stanzas before noting that apes "can't stand upright,/Unless to show fight,/With 'Du Chaillu,' that chivalrous knight!"

Years later it was revealed that "Monkeyana" was penned by Sir Philip Egerton, a geologist by training. It has become a regularly cited reference illustrating the early Darwinian debate. It was prompted not by the publication of *On the Origin of Species* but by the popular sensation stirred by Paul and his gorillas.

THE PRESS loved him—a "man of insignificant personal strength," according to the *Times*, who'd successfully slain a beast no one before had ever dared confront. Now he was being feted by elite Britain, an outsider who'd somehow wedged his foot in the door. Reporters delighted in the fish-out-of-water ambience that positively dripped from him.

According to one story, Samuel Wilberforce—still running victory laps after what he perceived as a triumph in his debate with Huxley—invited Paul to breakfast to meet about ten of the country's top men of science (or at least ten who'd earned the bishop's approval). When the appointed hour arrived, all the invitees gathered around the breakfast table— except Paul.

The next day, Wilberforce tracked down Paul on the street, curious to know why he'd been a no-show at the event hosted in his honor. A baffled Paul explained that he hadn't received the invitation from Wilberforce, a man so famous that he was known simply as "the Bishop" throughout London.

"But I left the note myself at your door," Wilberforce insisted.

"Must have miscarried," Paul replied. "I have seen no breakfast invitation except one from one 'Mr. Bishop,' and I make it a point to decline all invitations from people I don't know."

PARLOR MUSIC and dancing had become one of London's most popular pastimes. Someone would sit at the piano, while four couples would stand in a square formation, with each couple occupying a separate corner of the imaginary quadrangle. The pianist would play, and the dance—called a quadrille—would commence. The couples would take turns dancing in the center of the square, sometimes exchanging partners.

Publishing companies capitalized on the craze by selling sheet music that could accommodate this kind of dancing. The songs themselves became pop standards. In 1861, the stalls that sold sheet music stocked a new song, written by C. H. R. Marriott, titled "The Gorilla Quadrille."

The cover showed a drawing of a gorilla—dressed in tails and a bow tie—conducting an orchestra, while couples danced in the background. The music itself pulsed with primitive rhythms.

The lyrics were breezy and crude:

My name it is gorilla, and by that you plainly see
By birth I am a Darkie, but you can't get hold of me.
I laugh—Ah, Ah!

I sing—Doo Dah, Ah, Ah!
Doo Dah! Ah, Ah!
I'm th' wonderful gor-il-la who you've heard of but not seen.

ONE WEEK after *Punch* published "Monkeyana," the
magazine hit the newsstands with another cartoon featur-
ing London's favorite exotic species. This one was titled "The
Lion of the Season." The drawing featured an alarmed liveried
butler—eyes wide, hair standing on end—at a seasonal party
announcing the arrival of a guest: a gorilla, dressed in a tux-
edo and white gloves. The caption transcribed the startled
servant's introduction: "MR. G-G-G-O-O-O-RILLA!"

WITH LONDON transfixed by Paul's gorillas, a man
named Theodore Lent decided to visit the city on the Thames.

Lent's wife, Julia Pastrana, had died just over a year before,
in March 1860. She had spent her later years as a performer,
and Lent had been her manager. Together, they toured Germany, Austria, Poland, England, and Russia. Pastrana had
been born in Mexico with a rare condition that today is called
hypertrichosis terminalis, which means that her face and body
were covered with excessive amounts of straight black hair.
While traveling in Mexico, Lent had actually *purchased* Pastrana from a woman believed to be her mother. He dressed
her as a Spanish dancer and put her onstage as "the Bearded
and Hairy Lady."

Lent fathered a child with Pastrana, but the pregnancy
was riddled with complications. She gave birth to a son while
they were on tour in Moscow, and the boy was afflicted with

the same condition as his mother. He died two days later. Pastrana died three days after him.

Lent, however, wasn't ready to let go of his livelihood. He preserved the corpses of both his wife and his son.

In 1861, Lent arrived in London with the remains of his family. He placed them within glass cases. Both were positioned in stiff, upright postures.

"The figure," wrote the *Lancet*, in describing Pastrana's remains, "is attired in a dress of her own making, and it stands before us without the slightest odour, the slightest change, or the slightest appearance of corruption. The child is preserved with equal success in the same manner."

The author of the *Lancet* piece added that Pastrana's face appeared to be "the facsimile of the gorilla."

As he banged out the final chapters of *Great Expectations*, Charles Dickens continued to edit his weekly magazine, *All the Year Round*. Between serialized installments of his new novel, he published articles of general and literary interest. It was a demanding job, because Dickens believed every word printed in the magazine should be considered his own.

In the spring and summer of 1861, the novelist repeatedly indulged a recurring fascination with gorillas. Dickens would later write to Richard Owen, "If you knew how much interest it has awakened in me and how often it has set me a-thinking, you would consider me a more thankless beast than any gorilla that ever lived. But you do *not* know, and I am not going to tell you."

In the pages of *All the Year Round*, Dickens aired his obsession. His first of two articles about gorillas was both an appreciative summary of Paul's book and a meditation on what separates all things human from all things beastly. For Dickens, the difference seemed a matter of character development.

The magical transformation that had turned a gorilla into a man, the article stated, was perhaps the greatest miracle imaginable. Because humans were armed with intellect, morality, and a soul, the separation between mankind and apes was absolute, according to the magazine.

"The stupid weak savage will still make a prey of the yet more stupid but enormously more powerful gorilla, for the one uses reasons, and the other has only his instincts," the article stated.

As he grappled with the ending of *Great Expectations*, Dickens made sure his characters' fates were determined by their abilities to engage their intellects, moral sensibilities, and souls. He wrote two endings for his novel: in the first, his hero (Pip) and the girl he loves are kept apart, their love unrequited; in the second version, they exit the narrative hand in hand, presumably together forever.

Dickens chose the happy ending. Some critics would condemn his decision on the grounds of sentimentality, but the author stood firm, cynics be damned.

The novelist believed he knew exactly where such pessimists—Englishmen who "put a bad construction upon every word and deed"—should go: they should be thrown among the same deadly serpents, brutal savages, and vicious beasts that Paul Du Chaillu had encountered in Gabon.

"A gentleman of this disposition would enjoy himself to his entire satisfaction in the Gorilla Country," according to Dickens, "and the best thing he can do is to go there, and—to stay there."

R. M. BALLANTYNE had become the world's most popular writer of books for boys with the publication of *The Coral Island*—an adventure about three English youths shipwrecked on a Polynesian island. But by 1861, the success of that book, though just four years old, was already a fading memory. Ballantyne had since written three other books, but none sold as well. So he decided to write a sequel to the story that had made him famous.

Instead of another remote island, he set the new novel in the place boys now dreamed of being stuck: equatorial Africa.

In the book, the boys encounter elephants, hippos, biting ants, and slave traders. They hear fantastical legends about "hideous creatures one beholds when oppressed with nightmare." Eventually, the young heroes encounter the beasts, including one that "had broken Jack's rifle across, and twisted the barrel as if it had been merely a piece of wire."

The boys prevail in the end, earning a reputation among the natives as "the greatest hunters that had ever been born."

The sequel to *The Coral Island* was rushed into print in 1861 with the title *The Gorilla Hunters*.

LIKE HIS rival Charles Dickens, the novelist William Makepeace Thackeray edited a literary journal, the *Cornhill Magazine,* and in 1861 it was serializing chapters of his latest novel.

Thackeray, a satirist whose novels included *Vanity Fair* and *The Luck of Barry Lyndon*, saw the cartoon titled "Am I a

Man and a Brother?" in *Punch*, and he interpreted it as a veiled attack upon his magazine. "About the gorilla," he wrote in a letter to the co-founder of his magazine. "What do you think? That *Punch* picture is certainly against us."

Thackeray based his hunch, which was almost certainly wrong, on the fact that a recent installment of his novel had described a character with a "tea-spoonful of that dark blood." Thackeray had written that the character could expect to be rudely treated in America, but in England he'd be considered "a man and a brother." Somehow, Thackeray thought the editors of *Punch* were suggesting that *he* was an ape—a "Literary Gorilla," as he termed it in a satirical response printed in the *Cornhill Magazine*.

It might have been a case of literary paranoia, but Thackeray's suspicion made one thing clear: comparing someone to a gorilla had become the most fashionable and inflammatory insult of 1861.

LONDON'S POLICE courts were clogged with clusters of the city's unemployed, the down on their luck, the desperate.

On a day in mid-1861, a young woman was brought before one of the court's magistrates. She was accused of physically assaulting her little brother.

Her defense was one the court had never heard before. She justified the beating by explaining that her brother had called her "a gorilla."

FROM THE wings of the stage inside the Lyceum Theatre in the West End, a man named H. J. Byron watched nearly two thousand people fill the seats. Men and women were decked out in their best evening dress, in the mood for fun. Byron and the other members of the Savage Club—a bohemian social circle that comprised mostly journalists, playwrights, and actors—were eager to oblige.

The occasion was, officially at least, a fund-raiser for widows and orphans. But unofficially, it was the chance to see some of the most well-known satirists in London show off their wit and drollery in a farcical burlesque full of puns and topical jokes.

As Byron strode onto the stage to introduce the show, he wore a costume of thick fur underneath his top hat and long-tailed formal coat. It was as if Mr. Gorilla had stepped out of the pages of *Punch* and onto the Lyceum's stage.

"Behold me here!" Byron announced to the crowd. " 'The Lion of the Season.' Mr. Gorilla!"

In rhyming couplets, Byron explained that the stage-door keeper hadn't properly introduced him—because the man had run away in terror from the infamous creature of "Monsieur Chaillu."

> *Say, "Am not I a savage and a brother?"*
> *Do not I bear in this especial case*
> *A strong resemblance to the human race?*
> *Then let me hope, with pardonable vanity,*
> *To prove a link 'twixt our and your humanity.*
> *In brief—for sure I need no longer pause—*
> *In your good-will let me insert my claws;*
> *Spare not, I pray, your purses or your palms,*
> *The actors crave your hands, the fatherless your alms.*

———

THE ACTORS from the Savage Club weren't particularly creative in their choice of material. Every day of June on Regent Street, actors at the Grand Fancy Fair and Musical Festival performed *Gorumba and Little Billy*, a story about "Monster Gorillas" captured from the wilds of Africa and brought to England.

Exactly one day after that play ended its run, another show titled *Mr. Gorilla* opened at the fifteen-hundred-seat Adelphi Theatre. The playbill advertised Paul J. Bedford in the lead role of Paul Grandy, while an actor identified only as "An African Gentleman" had to settle for second billing in the title role.

AFTER WITNESSING the success that John Murray was having with *Explorations and Adventures*, Harper and Brothers published an American edition. Even though the country was in a civil war, it was published with much fanfare. According to some calculations, it was America's best-selling book of 1861.

An American critic in the *National Quarterly Review* chastised his countrymen for overlooking Paul while he'd been among them. He wrote of the "excellent opportunity wasted by them" when the young man had been in New York just a year earlier.

"And this man we allowed to pass by almost unnoticed," the reviewer lamented. "One or two interesting evenings at the rooms of the Geographical Society, two or three invitations to exceptional private houses, an almost deserted exhibition in a broken-down building on Broadway, and some cuts in *Harper's Weekly*, were our whole welcome to the man who has made all scientific Europe ring again and again with discussion over his discoveries."

The gorilla had become an international sensation, a symbol of a new and sometimes frightening era of discovery, and Paul had become an icon in his own right, swept up in a cultural tornado that soon carried him away.

⊰ ⏿ ⊱

CHAPTER 24

Three Motives

Even the most bleary-eyed patrons of the Elephant & Castle, stumbling out into the street from a fog of pipe smoke and ale stink, couldn't have helped noticing that this grimy corner of London was in the throes of a dramatic conversion.

The Surrey side of the Thames, a place where all the city's omnibus routes originated and the site of one of South London's busiest train depots, for years had been a drifters' hangout. The neighborhood's defining landmark had always been the pub, which had begun as an inn but was now one of the most notorious "gin palaces" in the city. Moral crusaders could often be found outside, temperance pamphlets in hand. But in the first half of 1861, directly across the street from the pub's swinging green doors, a new structure was rising out of the ground. This massive edifice of gray stone slowly grew to dominate the neighborhood. It was called the Metropolitan Tabernacle, and it was built specifically to accommodate the followers of a young Baptist preacher named Charles Haddon Spurgeon.

It was a new kind of cathedral, custom-made for a new kind of religion. With six stone columns out front, the exterior of the Greek-style building looked nothing like a traditional church. Inside, instead of rowed pews on the church floor, tiered balconies with iron filigree railings accommodated about five thousand seats, and standing-room sections could hold a thousand more people. There was no pulpit on the large oval stage—just a table, some chairs, and a couch. It looked less like a house of worship than like a concert hall. If it had been built a century and a half later, people might have called it a megachurch.

Traditionalists were aghast. They couldn't believe that Spurgeon charged five shillings for entry, as if he were some sort of song-and-dance man. Admission was strictly first come, first served.

"A monster place of worship, like all other monstrosities, is of a very doubtful propriety," observed the social critic Philip Cater; "indeed, monsters should not be permitted to live. . . . Those who have no shillings or sixpences to spare, will please to keep outside the building; for, according to the new order of things, the gospel is no longer without money or without price. It will be useless to make a rush at the doors without tickets, for the police will be stationed there to prevent all unqualified sinners from entering in!"

Spurgeon didn't care. He felt that charging admission not only raised money for his ministry but made people value his sermons more. He understood the dynamics of supply and demand, and he wasn't afraid to apply them to religion.

"Some persons, you know, will not go if they can get in easily," he explained to his followers, "but they will go if you tell them they cannot get in without a ticket."

His crowd was distinctly working-class, occupying "the social zone between the mechanic and the successful but not fashionable tradesman," according to one early observer. From the week the doors opened in March 1861, Spurgeon almost always attracted a full house.

Spurgeon roamed the tabernacle's stage like a prizefighter, and he had the nicknames to match: the Prince of Preachers, the Soul-Winner, the Son of Thunder.

He'd started out as a sort of novelty act, a boy preacher who spread the gospel at revivals across the English countryside, reared by a father and grandfather who'd been preachers themselves. By the time he was a teenager, he'd already become the pastor of his own church, the New Park Street Chapel. He was an established phenomenon—the voice of a new evangelicalism.

"I was told, and I believe, that in Agricultural Hall, in London, a place described as being like an unenclosed space for vastness, he made himself distinctly audible to 12,000 people," reported William Cleaver Wilkinson, a contemporary who wrote an introduction to Spurgeon's first biography. "It is even credibly affirmed that in the Crystal Palace, at Sydenham, he spoke, and was everywhere perfectly audible, to an assembly of 20,000 people."

In those days before microphones, Spurgeon guarded his voice as carefully as an opera singer. Backstage at the tabernacle, he'd sip chili vinegar with water, and he spiced his tea "as strong with pepper as can be borne" to keep his vocal cords in shape.

"Let us commence the present service by offering up a word of prayer!" With these words, he began his Sunday services that would attract an average of five thousand people each Sunday for thirty-one years.

At the exact same time that Charles Spurgeon became England's first superstar evangelist, he got his hands on Paul's book.

He devoured it, reading night and day from the first page to the last. As a traditional Christian, he considered the theory of evolution heresy, and the book's descriptions of an animal that was supposedly man's nearest animal relation were of an occupational interest. But the passages that stood out for Spurgeon were the ones in which Paul came across as a shining Christian example among the tribal heathens. Spurgeon took special note of this description of the adventurer among native Africans:

> *December 19th was Sunday by my account. I sat in my hut and read the Bible, and a great crowd came around and watched me with wondering eyes. I explained to them that when I read it, it was as though God talked with me. Then, to gratify them, I read aloud, and afterward tried to explain to them something of the teachings of Christ.*

Spurgeon saw Paul not simply as an explorer and hunter but as a torchbearer bringing the light of God to the Dark Continent.

Shortly after he finished Paul's book, a member of Spurgeon's congregation suggested that since all of England seemed to be obsessed with gorillas, why didn't the reverend address the topic at the tabernacle? The man, a painter, offered to provide Spurgeon with whatever visual props he might need.

"Let me paint a set of slides on the gorilla, and you give us a lecture," he told Spurgeon.

"Very well," Spurgeon said. "I will do it."

He knew that the traditionalists would find it uncouth, to say the least, for a man of the cloth to speak about gorillas in a house of God. But the subject would, Spurgeon knew, draw an awful lot of people to his church's door.

ABOUT 150 miles to the north, a Georgian mansion called Walton Hall sat on an island in the middle of a lake. It was the home of one of the most eccentric men in a country that claimed no shortage of them.

To reach Walton Hall's front entrance, visitors crossed a cast-iron bridge, then followed a stone footpath. Two large bronze knockers molded into the shapes of human faces stared out from the door. One was smiling. The other was contorted in an agonized grimace. The faces representing mirth and misery offered visitors the opportunity to choose a preference. If the visitor reached for the smiling face, he'd struggle for a moment before realizing the joke was on him: the rapper was welded down. That left no choice but to pick misery.

For the uninitiated, entering the vestibule was like stumbling into a nightmare. The sculpture of a mongrel incubus leered overhead—a baroque beast with the face of a man, the horns of a demon, the tusks of a wild boar, the ears of an elephant, and the wings of a bat. One leg ended in a cloven hoof and the other in a talon, which clutched a luckless serpent. The Latin inscription on the sculpture read, *Assidens praecordiis pavore somnos auferam*: "Sitting on the region of the heart, I take away sleep by fear."

With each step deeper into the house, the place got stranger.

Past the entrance to the dining room hulked the grand staircase, which led up through a dreamlike menagerie of dead creatures stuffed into vague approximations of their living forms. At the foot of the stairs sat something labeled *lusus naturae*, or "freak of nature": it appeared to be a sheep's head, but the horn was protruding from its ear. Each step of the staircase offered a new spectacle of the bizarre: carnival-colored birds from Brazil, monstrous lizards from Africa, an enormous boa constrictor. At the top of the staircase was a caiman—a crocodilian species native to Central and South America—that measured more than ten feet from snout to tail. Legend had it that the owner of this house had wrestled the creature to the death in the jungles of South America.

But that was when he had been a much younger man.

Now, in the summer of 1861, he was seventy-nine years old. But he hadn't lost any of the quirkiness that had defined his character since boyhood.

His name was Charles Waterton, but almost everyone knew him simply as the Squire. Various friends who struggled for the words to describe his physical aspect came up with "a person recently discharged from prison" or, more evocatively, "a spider after a long winter." He accentuated the look by squeezing into a brass-buttoned, swallow-tailed coat that was a couple of sizes too small and a couple of decades too old. He was shriveled, desiccated, worn—but far from feeble, and he loved to prove it. As a parlor trick, he'd scratch the back of his head with the toe of his right foot; then, from a cross-legged position on the floor, he'd rise to his full height without placing his hands on the ground. Sometimes, when walking along the shore of the lake with friends, he'd hop on one foot, just for fun.

But the Squire had always been plagued by one particularly irksome physical malady: pulmonary congestion. Every day, he rose at 3:30 a.m. to perform a ritual to relieve its symptoms.

He would get up from the bare plank floor where he always slept and where a beech-wood block served as his pillow. Moving to a chair, he sat and balanced a bowl on his lap.

He raised a cord of string to his mouth, holding one end in his teeth and using a hand to tie the length of it around his bare arm. He tied the string tight around his stringy bicep. A vein rose within the inner crook of his elbow. Discolored scar tissue shone on the creased skin. He then pressed the gleaming blade of a lancet to the vein.

He sat up straight and tilted his head back, drawing deep breaths as the blood began to trickle darkly into the bowl. As the fluid drained from his body, sometimes a full pint of it, his breaths seemed to flow more easily. Only after this ritual, which he called "tapping the claret," was he ready to start his day. And in the summer of 1861, those days were dominated by one activity: indulging an obsessive preoccupation with the young adventurer who'd become the talk of London.

Once, long ago, the Squire himself had garnered a lot of press as England's bravest jungle adventurer—the same title Paul had now seemingly captured. But Waterton didn't tolerate rivals easily. The new kid on the block made his blood boil. And Waterton wasn't alone in this sentiment.

JOHN EDWARD GRAY hadn't attended Paul's lectures at the RGS, but he wasn't about to miss the chance of seeing the gorillas in the display rooms at 15 Whitehall Place.

The rest of the city might have been blinded by the supposed grandeur of those specimens, but Gray eyed them like a jeweler who'd been cheated once too often. His gut told him they were counterfeits destined to lose their luster somewhere between first glance and second thought.

The fact that the Zoological Society—the organization that Gray chaired—hadn't been consulted before the gorillas were put on display was a slap in the face. Worse, Gray couldn't have helped but notice the adulation the British elite—the same ones who'd embittered him with rejections—were now heaping upon this young adventurer, who had zero previous scientific credentials. Du Chaillu's rise had depended on patronage, and it came from two people who would have occupied spots at the very top of Gray's "unfriends" list: Owen and Murchison.

Paul had been placed atop a very high pedestal. Gray couldn't resist trying to knock him off.

CHAPTER 25

The Gorilla War

After Gray toured the display room at 15 Whitehall Place, he got a copy of Paul's book and subjected the text to a kind of scrutiny that he was uniquely primed to provide: critical, exhaustive, and absolutely unforgiving.

His first salvo against Paul was published as a response to a generally favorable book review that had appeared in the *Athenaeum*. Gray began his letter by declaring that the "public seem to be under a delusion," and he proceeded to set the matter straight by venting his engorged spleen:

> *Some time ago the arrival of a new African traveller was announced. He read his paper at the Royal Geographical Society. It was soon discovered that his qualifications as a traveller were of the slightest description; but some of the Fellows seem to have been so taken with his tales about Gorillas and other animals, that they have allowed him to make one of their rooms into a museum and thus a great éclat has been given to his labours, certainly not on account of his geographical discoveries, for the map appended to his work is one of the most primitive that I have seen for years. If the Royal Geographical Society had transmitted the zoological*

notes and the collection to the Zoological Society, it would
have been seen that his qualifications as a naturalist were of
the lowest order, and that he has made few, if any, additions
to our previous knowledge.

Gray implied that Paul might have collected his specimens from natives on the coast of Africa, without penetrating
the interior at all. The methods of taxidermy employed by "the
traveller" (a derisive label that ranked below "the explorer" and
miles beneath "the scientist" in Gray's view) suggested that
"they have been preserved in or near the habitation of civilized men, and not in 'the forest' where 'daylight is almost shut
out.' " Even if they had been adequately preserved, Gray said
the gorilla skins *still* wouldn't have been worth the attention
they were getting because Europe had received its first gorilla
skeleton nearly five years earlier. He ridiculed Paul's "improbable stories" and noted that some of the illustrations included
in the book appeared to have been copied, without acknowledgment, from other sources. To underscore how undeserving
Paul was of England's reverence, Gray stated that among the
specimens labeled "new and undescribed" was a hoofed quad-
ruped that Paul had called a "white-fronted hog"; Gray had
recognized it as the same kind of African pig that was currently on display and *living* at London's Zoological Gardens.

"We are overburdened with useless synonyms," Gray
concluded, "and Natural History may be converted into a
romance rather than a science by travellers' tales, if they are
not exposed at the time."

Gray's criticism of the book's crude map could be
answered fairly easily. Paul himself had acknowledged during
his lectures that he had lacked the proper scientific measuring
instruments and admitted that his cartography was flawed.

But the other accusations were embarrassingly hard to counter. The illustrations included in his book had been prepared by artists in the United States, hired when Paul struck his original book deal with Harper and Brothers. In a few cases, including the drawing of the gorilla in the book's foldout frontispiece, the American artists had copied the illustrations from European journals. This brand of plagiarism wasn't illegal, because there were no copyright agreements in those days between America and Europe. Yet the accusations of cribbing stung Paul, who had earlier claimed that most of the illustrations had been drawn according to his own rough sketches.

As for his "new and undescribed" specimens, Paul insisted he had acted in good faith. In a letter to the *Times* printed four days after Gray's accusations appeared, he wrote:

> *I hope that neither in my book nor in my lectures I have pretended to be infallible as a naturalist, artist or traveller; yet I maintain that I have discovered in Equatorial Africa the new mammals and birds given as such in the list at the end of my volume. All of these were described in the published proceedings of two of the most scientific societies in America (with which Mr. Gray ought to be acquainted), some of the birds as far back as 1855, and I defy him to produce specimens existing in any European museum before that time.*
>
> *My map, at which he sneers, is a mere sketch map, it is true, but it was carefully prepared from observations made on the spot with the compass, and I will vouch for its general accuracy.*
>
> *My illustrations, prepared, not in this country, as he asserts, but in America, were taken either from my own rough sketches or from the actual objects, with the exception of four or five out of a total of 74.*

*Would it not have been more fair of Mr. Gray, before
giving vent to insinuations that I had never visited the
countries which I describe, nor collected in those countries my
natural history specimens, to have applied to my friends at
Corisco and on the Gaboon, whose names are mentioned in
my book? Mr. Gray pretends to be in communication with
the missionaries and traders in those parts, and therefore this
course would have been the more obvious, as he would have
saved himself from the imputation of uttering mere calumnies.*

I am, Sir, your obedient servant,

P. B. Du Chaillu

Paul was more forceful in another letter he sent to the *Athenaeum*. He painted Gray as a pampered academic whose work depended on others brave enough to go where he wouldn't dare tread.

"This at least is certain," he wrote, "that the naturalist who works at home, safely and luxuriously lodged in his museum, has now, through my travels in African forests, the opportunity of acquiring the knowledge of the species. The return which Dr. Gray makes me, reminds me of the ape that grins a malicious snarl at the hand that has just given it a dainty."

Gray wasn't going to let Paul have the last word. He fired off another letter to the *Times*, with more incriminating evidence. He said that the ape that Paul called a nest-building "nshiego-mbouvé" in both his book and his exhibition was a common chimpanzee. Paul had, in fact, erred; he mislabeled it a separate species based on the fact that it lacked hair on the top of its head and lived in a shelter built in trees. In fact, baldness can occur in all subspecies of chimpanzee, and all can build nests. The mistake was helpfully exposed by his pirating illustrators. They had clearly based their drawing of the nshiego-mbouvé on a photograph of a common chimpanzee that had been kept in the Jardin des Plantes in Paris.

Gray had Paul pinned. He knew the young man's only defense was to plead ignorance.

"If Mr. Du Chaillu had published his work as the 'Adventures of the Gorilla Slayer' I should have taken no notice of it, for readers of such works like them seasoned to their palate," Gray wrote. "It is only as the work of a professedly scientific traveller and naturalist that I ventured any observations on it."

Another of the "new" species Gray cast doubt upon was an otter-like animal that Paul had dubbed the *Potamogale velox*. Based on the partial specimen of the animal Paul had collected, Gray claimed that the adventurer's classification of the animal was riddled with error. He suggested that Paul's genus and species classifications be scrapped. Gray's allies in the press proposed a substitution: the *Mythomys velox*, to reflect the mythmaking propensities of its discoverer.

AND WITH that, mistrust ravaged Paul's credibility. Every statement of fact in his book was now vulnerable to a contagion of doubt.

Heinrich Barth, a famous German explorer whose travels helped map much of northern Africa, amplified Gray's attacks on Paul's crude map. In a scathing review published in Germany's most prestigious geographical journal, Barth said that the fact that Paul traveled without a scientific instrument of any kind nullified his claim to the title of "explorer." What's more, several of the few dates listed in the narrative seemed contradictory: letters published in the journal of the Philadelphia academy suggested that Paul had been on the Atlantic coast

A male western lowland gorilla, the species that Du Chaillu encountered in Gabon. *Volodymyr Burdiak*

Du Chaillu poses in the same kind of outfit he wore during his expeditions for this portrait, taken in 1861. His slight build and youthful appearance made him, according to one magazine, "almost the last man whom one at first sight would set down as a great explorer, adventurous traveler, and a naturalist of no ordinary attainments." *Courtesy of the Royal Geographical Society*

Richard Owen, who became Du Chaillu's de facto patron, is now remembered mostly for his opposition to Darwin's theory of natural selection. But he was also his generation's most celebrated anatomist and a target of jealousy among those who envied his cozy relationships with England's most powerful leaders. *Courtesy of the Miriam and Ira D. Wallach Division of Art, Prints, and Photographs, New York Public Library, Astor, Lenox, and Tilden Foundations*

(Below) John Edward Gray was one of the leading zoologists of his era—and also one of the most combative, unafraid to enter public disputes with his peers. *Courtesy of the National Library of Medicine*

(Above) Thomas Henry Huxley, "Darwin's Bulldog," positioned himself as the most energetic early defender of the theory of natural selection. "I am prepared to go to the Stake if requisite," he wrote to Darwin when *On the Origin of Species* was published. *Courtesy of the Miriam and Ira D. Wallach Division of Art, Prints, and Photographs, New York Public Library, Astor, Lenox, and Tilden Foundations*

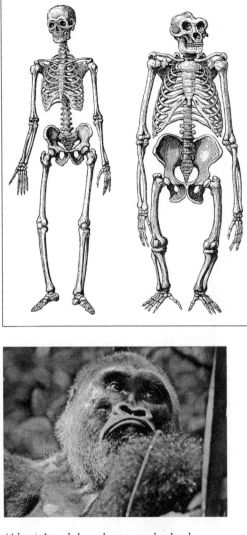

By comparing the features of the human skeleton *(left)* with the gorilla skeleton *(right)*, Richard Owen reasoned that the gorilla was man's closest relative in the animal kingdom—a conclusion that turned the newly discovered gorilla into a symbol of the evolution debate. Molecular analysis has shown that chimpanzees have slightly more similarities to man than gorillas do on a genetic level. *Vintage illustration from* Die Frau als Hausärztin, *1911*

(Above) An adult male western lowland gorilla. *Uryadnikov Sergey*

(Right, above) A female western lowland gorilla. *Ronald van der Beek*

(Right, below) A male western lowland gorilla is shown in his native habitat. *Uryadnikov Sergey*

Du Chaillu's gorillas sparked a craze in 1861 for everything ape-related. Enterprising authors, playwrights, and songwriters cashed in on the phenomenon, producing ephemeral hits like "The Gorilla Quadrille." *Courtesy of the National Library of Australia*

At the same time Du Chaillu displayed his gorillas on Broadway, P. T. Barnum unveiled a living exhibit he called the "What Is It?" Barnum claimed the "creature" had been captured in Africa, but in truth he'd clothed a man suffering from a disfiguring developmental disorder in a suit of fur. *Courtesy of the New-York Historical Society*

Sir Richard Burton (shown here late in life) befriended Du Chaillu and followed his footsteps by searching for gorillas in Gabon. *Courtesy of the Library of Congress*

(Below) As president of the rapidly expanding Royal Geographical Society, Roderick Impey Murchison sponsored British explorers all over the globe. Most famous among them were his "lions"—travelers like David Livingstone—who ventured to unmapped regions of Africa. *Courtesy of the Library of Congress*

(Above) The Reverend Charles Haddon Spurgeon, the most popular evangelist in Victorian England, opened his Metropolitan Tabernacle weeks after Du Chaillu's gorillas debuted in London. Within weeks the Prince of Preachers was addressing the subject of evolution, inviting Du Chaillu—and one of his specimens—into his tabernacle.
Courtesy of the Library of Congress

(Above, left) This cartoon, called "Monkeyana," appeared in *Punch* in May 1861. The gorilla's placard parodies the slogan used by slavery abolitionists in England.

(Above, right) The frontispiece of Du Chaillu's bestselling *Explorations and Adventures in Equatorial Africa* (1861) depicted a male gorilla in a humanlike pose, complete with strategically placed foliage to shield readers from the animal's immodesty. *From* Explorations and Adventures in Equatorial Africa *(1861)*

Du Chaillu attempted to take photographs of gorillas (such as the ones comically depicted here, captured during his second expedition), but the plates did not survive his rushed retreat. *From Du Chaillu's book* The Country of the Dwarfs *(1872)*

One of the most thrilling—and doubted—stories Du Chaillu told was of
a native hunter who'd been attacked by a gorilla and left for dead with the barrel
of his gun bent, presumably by the gorilla. *From* Explorations and Adventures in
Equatorial Africa *(1861)*

An illustration depicts Paul and his African porters fighting back against the
villagers they had angered—and infected with smallpox—during their journeys.
From Du Chaillu's book Journey to Ashango-Land *(1867)*

After being criticized for not using astronomical observations to verify the route of his travels, Du Chaillu apprenticed with some of England's leading astronomers and geographers, learning how to use sextants and chronometers to determine his latitude and longitude. *From Du Chaillu's book* The Country of the Dwarfs *(1872)*

Du Chaillu, pictured here in the 1890s, spent most of his later years in the United States, which he often claimed as his homeland. "He still speaks with a decided French accent that betrays his nativity in the old city of New Orleans," the *Washington Post* reported near the end of his life. *Courtesy of the Harvard Fine Arts Library*

at the same time that in his book he claimed to have been in the interior. Barth concluded that Paul "deliberately falsified material" and "has forged at least a great part of his travels." Barth suggested that the young man didn't venture more than a few miles away from the coast, and the rivers and mountains he described deserved no geographical credence. The German assailed what many considered Paul's most important geographical contribution: his conclusion that the Ogowé River was part of a large system that included important tributaries, such as the Ngounie River, which he'd been the first to describe. Barth deemed these "discoveries" worthless; in fact, he doubted that Paul had ever laid eyes on those rivers.

When a new map of equatorial Africa was published in Germany after Barth's damning verdict, geographers listed the names of several of the inland villages that Paul had claimed to visit—but placed them near the Atlantic coast.

EARLY IN his book, Paul had written how he exploited his reliable marksmanship to impress the natives. "As we were lazily sailing along, I espied two eagles sitting on some high trees about eighty yards off," he wrote. "Willing to give the fellows a taste of my quality, I called their attention to the birds, and then brought both down with my double-barrel." But shortly after Gray's letters appeared in the press, a rumor began circulating—first by word of mouth, then in the press—that Paul had been spotted at the Wimbledon Common, where the National Rifle Association maintained firing ranges. According to the story, someone asked him to take a turn with a long-range "ball" rifle, and Paul was forced to awkwardly reply that he had never used that sort of gun. Hearing the story, skeptics concluded that Paul must have used a common "fowling gun" in Africa. That sort of gun had a range of accuracy limited to about twenty yards. Viewed in this light, Paul's stories of picking off eagles in high branches seemed ludicrous.

Yet there was a significant flaw in the rumor: Paul had never set foot inside Wimbledon Common. He was, furthermore, a genuinely accomplished shot with a ball rifle. That didn't stop the gossip from further eroding his credibility. Paul was suddenly famous *and* infamous, a modern Munchausen, a mere spinner of yarns. Those yarns seemed to be unraveling in a messy tangle.

The man and his gorillas had united London in wonder, but now they were forcing people to take sides in what the press began to label "the Gorilla War."

EVER SINCE Roderick Murchison had thrown the successful farewell banquet for David Livingstone at the Freemasons' Hall, he regularly booked the venue when he needed a little bit of pomp and a lot of space to promote the Royal Geographical Society. To celebrate the organization's anniversary in 1861, he chose the tavern for a Monday evening banquet that promised to stretch into the wee hours of Tuesday.

It had already been a long day for Murchison, who'd overseen the anniversary meeting of the group that afternoon at Burlington House. He'd presented the Founder's Medal to John Hanning Speke, who after splitting with his co-explorer Richard Burton had recently discovered Lake Victoria, believed to be the long-sought source of the Nile River. But Murchison had spent much of that meeting clarifying exactly where he and his organization stood when it came to the Gorilla War. Paul, seated in the audience at Burlington

House, was putting on a brave face despite the public ridicule. His spirit seemed as buoyant as ever—constantly besieged, but continually popping up like a gourd in rough water.

He had been busy writing an updated introduction for a new edition of his book, and he was confident that it would calm his harshest critics. He told people that he had found one misprint concerning a date that he hadn't noticed; however, that hadn't been the cause of his conflicting chronology. In the first edition, he said, he had grouped his expeditions of the areas north of the Gabon River together in successive chapters. In reality, those northern expeditions had been interrupted by other journeys to the southern regions, which likewise were grouped together in the narrative. Livingstone had similarly organized his book's narrative according to geography, to help reduce confusion on the part of the reader. In his new introduction, Paul admitted that he should have made it clear he had used this technique in his original preface. To make amends, he inserted into the new edition a chronological table of his journeys, listing the specific dates that he visited each place described in his book.

His efforts put his friends at the Royal Geographical Society at ease. At the anniversary meeting in the afternoon, Murchison wanted him to know that he was among friends. At the rostrum, Murchison pronounced that not only were Paul's explorations trustworthy, but they represented "one of the boldest ventures which man ever undertook." The audience burst out in applause.

"Strikingly attractive and wonderful as were his descriptions, they all carry in themselves an impress of substantial truthfulness," Murchison continued. "Of this no one who has formed the acquaintance of M. Du Chaillu, and looked into his open countenance and met his bright and piercing eye, can for a moment doubt."

If Paul had been comforted by Murchison, he must have been overjoyed at the tavern later that evening. The party seemed less an anniversary celebration than a concerted effort to buck up the spirits of the embattled young adventurer.

At the head table sat knights, dukes, earls, lords, MPs, counts, aldermen, the mayor of London, and, of course, Richard Owen. After everyone had tucked into their food, Owen rose, lifted his glass, and proposed a toast: to the health of Paul Du Chaillu.

On the spot, Owen attempted to refute Gray's contentions, one by one. The gorilla skins Gray had inspected had clearly been prepared with arsenic near where they were killed—not on the coast. The privations Paul had endured were unimaginable, the dangers were horrific, and the fruits of his labors were a boon to science.

"Whether one judges Monsieur Du Chaillu by personal discourse, by his material evidences, by what he appeared to have seen of the living habits of the animals he described—testing those accounts by what we know of their structure—or by the incidents and style of his narrative, he impresses one with the conviction that he is a truthful and spirited man of honor and a gentleman," Owen said, raising his glass to the audience. "I have much pleasure in proposing his good health."

A hearty round of "Hear! Hear!" echoed through the hall, and a call went up for Paul to say a few words.

He stepped to the front of the hall, enjoying the warm rush of goodwill.

"I feel almost overwhelmed by the compliment that has been paid to me," he began, "the more so because I have been the object of a bitter attack—I don't know why—but relying on the truth of what I have written, I knew that in this noble-hearted country there were men who would do me justice."

The audience cheered loudly, but Paul had more to say.

"If I had been in my own country, these attacks would have been rebutted by friends who knew me from my boyhood, and who knew I am incapable of being an imposter."

With the unshakable good cheer that had charmed so many in London, he insisted that he bore no personal grudge against Gray. The truth, he said, would prevail in the end.

"I am just a boy," he said, "and the more I come into contact with the great men in this and other countries, the more I am convinced that they will not see me crushed."

GRAY WASN'T finished yet, and the reports of the banquet at the Freemasons' Hall seemed to fuel his determination to unmask Paul as a fraud.

About eighteen hours after the party at the tavern, Gray strode into a meeting of the Zoological Society. He was ready to take his attack to a new level. Instead of referring to him as a mere traveler, now Gray dismissed Paul as "an uneducated collector of animal skins for sale and an exhibitor of them in Broadway, New York." That description included a veiled dig at Owen, who had agreed to purchase some of the skins on behalf of the British Museum, where they would be displayed—without Gray's consent.

When Paul had lashed out defensively, painting Gray as an armchair naturalist, he ignited the museum keeper's instinct to go for the jugular. Gray went ballistic—literally.

He reported that a "friend" of his examined the skins and skeletons on display at Whitehall Place, searching in vain for visible bullet holes. "He says they seem to have been wounded when retreating, and not attacking," Gray said.

Gray then visited a taxidermist who had prepared one of

the skins that would soon be displayed at the British Museum. It was a large male gorilla, like the one Paul had described as charging close to him before he shot it in the chest. "I then inquired of [the taxidermist] whether he had observed any bullet hole in the chest, and he stated that he had not, but pointed out to me two holes in the nape of the neck (now filled with putty)," Gray explained. "There are also two large holes in the thin portion of the hinder part of the skull belonging to the same skin which pass through the bone, and are quite sufficient to have caused death."

Despite the authority with which Gray voiced his contentions, they weren't conclusive. Sir Philip Egerton, a well-known sportsman (and the anonymous author of the "Monkeyana" poem that appeared in *Punch*), had examined the same specimen that Gray referenced *before* it had been stuffed. Egerton insisted the evidence was entirely consistent with a frontal shot. Additional experts concluded that the holes in the back of the gorilla's skull couldn't have come from bullets, as Gray had implied. That part of the bone was paper-thin—about two-thirds of a millimeter. Those holes likely appeared after the animal's death, caused by rough handling.

Though Owen and Huxley still sparred over the gorilla's place in the evolution debate, more of the attacks against Paul were personal. At least in this particular instance, Gray's argument appeared to be based less on verifiable evidence than on a grudge. The controversy was shaping into a long war that would rage far longer—and range far wider—than Paul could have predicted.

·◦⫴◦·

CHAPTER 26

The Squire's Gambit

Charles Waterton wouldn't have been able to read about Paul without recognizing something familiar about the story. Years before, the Squire's honesty had been questioned during a remarkably similar public controversy concerning his own jungle adventures.

At twenty-two Waterton first traveled to British Guiana, at the northern edge of South America, to oversee an uncle's property. It wasn't long before he set out into the wilderness, dipping south into a forest that blanketed the northern third of the continent. For nearly a dozen years he continued to explore the South American jungle as far south as Brazil, generally traveling on foot, often without shoes. He took copious notes on the region's flora and fauna, transforming himself into a competent and observant naturalist. In 1825 he published a book called *Wanderings in South America*.

It was an unexpected success. Members of an upcoming generation of roving naturalists—Charles Darwin and Alfred Russel Wallace among them—later cited the book as one of several that excited their interest in natural history and exploration. But critics smelled hyperbole.

In his book Waterton had written that he was always

on the lookout for the crocodile-like caiman, hoping to add a flawless specimen to his collection. He'd wade into the swampy haunts of the reptile, which was reputed to be a man-eater, but he'd come out empty-handed. Finally, with a small group of natives, he devised a plan to fish for a specimen. They tied a sharp, meat-laced hook onto the end of a long rope and cast it into a river overnight. The next morning, they found that they'd hooked one.

Waterton wrote that the frightened Indians wanted to shoot the caiman with arrows, but he begged them to put down their weapons. He wrote that he waded into the water and wrestled it ashore himself, riding the writhing creature onto dry land as one might ride a horse.

"I immediately seized his forelegs," Waterton wrote, "and by main force twisted them on his back; thus they served me for a bridle." The book's illustration was comically faithful to his description, depicting him calmly astride the scaly beast.

Those who knew him best recognized that wading into water with a caiman was perfectly in keeping with his slightly unhinged brand of bravery: they couldn't forget that he had once fashioned a homemade pair of wings, strapped them around his arms, and jumped off his roof, convinced that he'd be able to fly (he could not). But those unfamiliar with his eccentricities, or his tendency to embellish his stories for effect, accused him of inventing the episode out of whole cloth.

Waterton was changed by the controversy surrounding his book, but instead of being sympathetic to later genera-tions of adventurers, he tended to be reflexively hostile to anyone who challenged his status as the most famous living collector of exotic specimens. Waterton had fashioned himself as the original fair-skinned knight striding fearlessly among the quaking natives, an erudite naturalist who doubled as a man of action. The untraveled scientists in the professional societies—"self-constituted censorious scoundrels," as Waterton sneeringly called them—didn't deserve a fraction of his acclaim. In his later years, Waterton seemed hell-bent on protecting his claim as the king of wilderness adventurers.

Central to his image was Waterton's Roman Catholic faith (he claimed to be descended from no fewer than eight canonized saints). God had given man dominion over nature, Waterton believed, so anyone who feared nature was spiritually weak. He protected his image as a fearless wrestler of monsters because it placed him not *within* the natural world that he loved but *above* it.

But now, decades after the publication of Waterton's book, Paul was taking his place as the prototypical beast-slaying adventurer. The gun-toting gorilla hunter, not the unarmed Squire, had become the model inveterate thrill seeker who rarely failed to bag the big-game trophies he'd set out to conquer. This type of stock character would later be branded the Great White Hunter in the popular lexicon. The implied derision that would eventually stick to that label hadn't yet established itself, and England couldn't stop talking about Paul.

WATERTON HADN'T seen Paul's specimens, he hadn't read his book, and he hadn't heard him speak. But that didn't stop him from anonymously penning the very first article attacking Paul's credibility. Gray's articles, which captured far more public attention, actually appeared shortly thereafter. With Paul's reputation publicly tarnished by Gray, Waterton dropped all pretense of anonymity.

He published several critical pieces in the *Gardeners' Chronicle* and added some new accusations to Gray's. Believing that all apes were more or less similar, Waterton was convinced that gorillas should be found in trees—like the arboreal monkeys of the New World, with which Waterton was most familiar. Gorillas should not be spotted on the ground, where Paul had encountered them. He also doubted that gorillas could rise up from all fours to stand on two legs. Quoting from news reports of one of Paul's lectures, in which he described the gorilla as beating his chest, Waterton mocked the idea: "It must have been on its hind legs only; a most trying and unsteady position for an ape in the hour of battle!"

Waterton's successive articles throughout the summer of 1861 reiterated that he found it impossible to believe that a gorilla—or any ape—would charge a man. He mentioned that he himself had experienced numerous encounters with apes, and none were beyond his capacity to tame. "Our closet naturalists may gulp such foreign food as this, and praise its flavour: but I remove it disdainfully from my lips, as it ill befits them."

Waterton then casually referenced a fact intended to destroy Paul's credibility once and for all: *the Squire had once himself owned a live gorilla and for years had kept it as a playful, gentle pet.* The implication was clear: not only had Paul misrepresented his journeys and the nature of the animal he so famously encountered, but Paul didn't even deserve credit for bringing the best specimens to England. Waterton did.

BEFORE PAUL came along, Waterton had not been interested in gorillas. But by that summer, he was convinced not only that he had seen one years earlier than Paul but that he currently owned a stuffed gorilla and kept it on display in his house.

Among Waterton's hobbies was anthropomorphic taxidermy: he stuffed zoological specimens, dressed them, and positioned them as still-life actors in elaborate theatrical dioramas. Those displays customarily mocked all those who dared to challenge his beloved Catholic Church. In one scene shown under glass in Walton Hall in the summer of 1861, he arranged various reptiles in positions of attack around something labeled "The True Church." A toad was labeled "Martin Luther," and a snake was "John Calvin."

But the oddest diorama was titled "Martin Luther After the Fall." It centered on a small stuffed ape in a position of humiliating absurdity, complete with a set of donkey ears affixed to the top of the creature's head. Waterton had believed the animal was a young chimpanzee. But now he'd begun to reconsider the assumption. Abraham Bartlett, the taxidermist who had tried to salvage the rum-soaked gorilla skin that was sent to Owen a few years earlier, visited the mansion in 1861. Bartlett recognized the animal at once: a baby gorilla.

The young animal, named Jenny, had been captured by natives near the Congo River and brought to Europe in 1855 as part of a road show called Mrs. Wombwell's Travelling Menagerie. Waterton was fascinated by the tiny creature, and he convinced Mrs. Wombwell to let him play with Jenny on four separate occasions. Their final meeting, according to Waterton, was full of sweet sorrow. "Having mounted the steps which led up to her room, in order that I might take my leave of her, Jenny put her arms round my neck; she 'looked wistfully at me,' and then we both exchanged soft kisses, to the evident surprise and amusement of all the lookers-on," Waterton wrote. "'Farewell, poor little prisoner!' said I. 'I fear that this cold and gloomy atmosphere of ours will tend to shorten the days.' Jenny shook her head, seemingly to say, there is nothing here to suit me. The little room is far too hot, the clothes they force me to wear are insupportable, whilst the food which they give me is not like that upon which I used to feed when healthy and free in my own native woods. With this we parted—probably for ever."

His hunch was right. Jenny died within months. "She journeyed on, from place to place, in Mrs. Wombwell's fine menagerie of wild animals, till they reached the town of Warrington, in Lancashire," Waterton wrote. "There, without any previous symptoms of decay, Jenny fell sick and breathed her last." But the Squire had made arrangements with the show's caretaker, a woman named Miss Blight, to recover the ape's remains if misfortune struck. "Miss Blight wrapped her up in linen by way of winding-sheet, put her in a little trunk, and kindly forwarded her to Walton Hall, at the close of February, in the year 1856."

Putting aside any heartache over her death, Waterton dissected and stuffed Jenny, preparing her for the starring role in his museum of curiosities. Though he'd never been to central Africa himself, he assumed that the primates there were similar to the ones he'd observed in South America. Following this chain of logic, weak as it was, he decided that he was eminently qualified to speak authoritatively on the habits of African apes in the wild.

He remembered that Jenny liked to knuckle-walk, which Waterton thought was a painful form of locomotion that resulted from captivity. If the animal had been allowed to spend her time in the treetops, as was natural among the monkeys of South America, he assumed that Jenny's movements would appear more graceful. Gorillas, he concluded, were made to swing in trees, not walk on terra firma. To Waterton, the fact that Paul almost always described finding his gorillas on the ground was an obvious sign of the author's fraudulence. Another clue was his descriptions of them charging him: Jenny had been a gentle playmate, unlike the intimidating brutes that Paul had described. She never would have charged a soul.

In the early summer, Waterton traveled to London hoping to prove that Paul had invented his stories of untamable gorillas. No animal, he said, was beyond man's powers of control.

Waterton wanted to put on a show to prove his point, and he knew exactly how to do it. A few years earlier, he had visited London's Zoological Gardens to examine an orangutan. He had entered the animal's cage without incident, but almost as soon as he left, the ape urinated on the floor. Waterton wrote that he was "scandalized beyond measure, at this manifest want of good breeding," and he considered it firsthand proof that "all monkeys are infinitely below us—aye, infinitely indeed." Now Waterton intended to restage his foray into the orangutan's den. Assuming that gorillas were more or less the same as orangutans, he hoped his public one-on-one encounter with an ape would undermine Paul's reputation.

Richard Hobson, a friend of Waterton's who wrote a flattering, though often defensive, biography of him shortly after his death, painted the Squire as a conquering hero who proved his mastery over the ape with "cool courage":

I allude to an occurrence in the Zoological Gardens, in London, in the year 1861, when, after much entreaty on the part of Mr. Waterton, he was permitted by the then curator, Mr. Mitchell, now deceased, to pay his personal respects to a large orang-outang, from Borneo, which was reputed to be

*very savage. Indeed, the keepers, one and all, declared that
"he would worry the Squire, and make short work of it," if
he should enter his den, especially as he was just then in a
horrid temper, having been recently teased by some mischievous
boys. The late Mr. Mitchell, even at last, yielded to Mr.
Waterton's urgent request, with great reluctance. Nothing
daunted by all this badinage of the keepers, the Squire, to
the very great horror of numerous spectators, entered the
palisaded enclosure with a light heart. The meeting of these
two celebrities was clearly a case of "love at first sight," as
the strangers embraced each other most affectionately; nay,
they positively hugged each other, and in their apparently
uncontrollable joy, they kissed one another many times, to the
great amusement of the numerous spectators.*

Waterton was certain that he had exposed Paul's incompetence as an observer of nature. On July 14, he wrote a letter to a friend predicting that "experts" like Owen and Murchison—the same class of scientists who had criticized Waterton's book decades before—would regret their support of the young man.

"What a clever fellow Du Chaillu has been to have enlisted in his favour the powerful approbation of our learned doctors in zoology," Waterton wrote. "But whoever shall read my letters on his gorilla in the *Gardener's Chronicle* will, I trust, conclude that those gentlemen ought to have examined well the dangerous locality before they took the leap."

··◦❘❘◦··

CHAPTER 27

The Gorilla in the Pulpit

Charles Spurgeon followed the controversy that surrounded Paul, but according to his reading of *Explorations and Adventures* the young man was a good Christian who'd been the victim of misplaced outrage.

"It's nothing but the pictures," he told himself, noting how the illustrations seemed to threaten the author's credibility much more than the actual text did. "The pictures have ruined the book."

As he prepared to talk to his congregation about the gorilla, the preacher sought some visual props that would prove far more arresting than mere illustrations. He wanted one of Paul's gorillas, and Paul himself, in the flesh.

As SOON as the side doors of the Metropolitan Tabernacle opened on Tuesday evening, October 1, the mad scramble for seats began. Charles Spurgeon's special lecture—"The Gorilla and the Land He Inhabits"—had been oversold by the ticket office. Hundreds of people were turned away at the door, even though some of them had secured reservations days in advance.

A small army of ushers guided the crowd through the aisles, making sure not a single seat or standing-room space was left empty. Spurgeon, meanwhile, waited in the vestry backstage. He was in a good mood, greeting those around him with pumping handshakes. He stood about five feet six, but his stout frame lent him a weighty presence. His bearded cheeks had grown plump, and his hands swelled with gout. His heavy-lidded eyes narrowed to slits when he smiled.

On this night, he had good reason to smile as six men carried in a bulky parcel wrapped in coarse linen. At the front of the stage they began to unwrap it. They fastened one of the beast's hands to the iron railing that encircled the stage and raised its other arm overhead, pointing toward the audience. The congregation howled with laughter, recognizing the gesture: it appeared as if the gorilla were delivering a fiery sermon—just like Spurgeon.

After a few minutes, Austen Layard, the Parliament member who represented the tabernacle's South London district, walked to the middle of the stage. Paul sat in one of the velvet-lined chairs alongside Spurgeon at the side of the altar.

"I'm at a loss to know why I've found myself here," Layard told them—an unconvincing confession, considering that Spurgeon had given him the irresistible opportunity of face time in front of thousands of people, some of whom almost certainly were eligible voters. "It surely isn't to introduce the reverend lecturer, who must be known to all or the greater part of you personally—if not personally, then there are few throughout the length and breadth of the land, whether rich or poor, high or low, to whom his name is not familiar."

Layard looked to his right, where the gorilla stood in his attitude of frozen pontification. "We are now to be entertained by Mr. Spurgeon's lecture on the gorilla," he said. "But in after ages, according to the 'development theory,' we shall doubt-less have a gorilla lecturing on Mr. Spurgeon!"

To win the favor of a religious-minded constituency, the politician mocked the scientists. The strategy, Layard discovered, worked very well: the crowd erupted in laughter. The portly reverend rose, as if on cue, and approached the center of the stage.

"Mr. Chairman, and my very good friends, I am very glad to see you here, though you have taken me very much by surprise," Spurgeon said. "I was reckoning upon a quiet evening with a moderate audience, but you have crowded this vast house, and I regret to say there have been great multitudes turned away from the doors. We are doomed to disappointments, but such as these one can afford to endure with equanimity."

He was in fine form. He fed on the audience's laughter, and he encouraged it by assuring them there was no harm in letting loose with a guffaw or two. Christians needed an outlet for entertainment, he believed, and what better place to find it than in a church? The tabernacle shouldn't be restricted to dry theological discourse, he told them. Entertainment, science, politics—his church had room for *everything*.

With that, Spurgeon turned toward Paul.

"So, is Mr. Du Chaillu's book true or not?" Spurgeon asked rhetorically. "You can see him for yourself."

Thousands sized up the unimposing adventurer, who remained at the side of the stage.

"When you look at him, you would hardly think he was able to shoot a gorilla, or to bring home as many as twenty-two."

Thousands exploded in laughter.

"I do verily believe, in spite of all that has been said, that

Mr. Du Chaillu's book is matter of fact. It is not written so carefully as a scientific man might write it, nor so orderly and regularly as the author might re-write it, if he had another seven years to do it in. Yet I believe that it is true, and that he himself is worthy of our praise as one of the greatest modern discoverers—a man who has done and dared more for science, and, I think I may add, more for the future spread of religion, than most men of his time."

Spurgeon quickly turned to face the gorilla.

"He is an enormous ape, which claims to approach the nearest to man of any other creature," he said. "How nearly he approaches, I leave you to judge."

Spurgeon was steering the talk toward evolutionary theory. He wanted to make certain that his congregation knew which side of the debate he endorsed.

"If we should admit this gentleman to be our cousin," Spurgeon declared, wheeling away from the gorilla and back toward his audience, "there is Mr. Darwin, who at once is prepared to prove that our great-grandfather's father—keep on for about a millennium or two—was a guinea-pig, and that we were ourselves originally descended from oysters, or seaweeds, or starfishes!"

He was working himself into a lather, and the crowd rewarded him with applause. "Seriously, let us see to what depths men will descend in order to cast a slur upon the Book of God. It is too hard a thing to believe that God made man in His own image; but, indeed, it is philosophical to hold that man is made in the image of a brute, and is the offspring of 'laws of development.' Oh, infidelity!"

Du Chaillu listened in silence as his gorilla became the subject of a two-hour sermon that ended with Spurgeon stressing the importance of missions in Africa.

Spurgeon concluded, "I am pleased to say that my friend Mr. Du Chaillu—if he will allow me to call him so—wherever he has been, has sought to open the way for missionary efforts, and has been the missionaries' friend everywhere."

Paul had, in fact, done precisely nothing to pave the way for missions during his travels. He could have proudly accepted the title that Spurgeon seemed so eager to bestow— that of the voyaging religious crusader—but Paul didn't. He thanked Spurgeon for defending the general truth of his book and for underscoring its faults in a good-hearted spirit.

"I can only say that if I travel again, I will try and do better," Paul said. "I have learned a great deal of wisdom during the last five or six months, and I will put it into practice during my next travels."

Spurgeon announced that all the proceeds from the night's event would go to the Band of Hope Union, an organization promoting total abstinence from alcohol—a cause that Paul had dismissed in his book, declaring that a daily tipple of wine, brandy, or ale was "absolutely necessary" to maintain a traveler's physical equilibrium in Africa. A children's chorale group brought the evening to a close, singing gospel hymns as the audience began filing out of the tabernacle. One writer attending the event described Paul's reaction to the entire spectacle as "bewildered."

IF PAUL wasn't sure what to make of what he'd witnessed at that altar, London's cultural critics knew exactly how to assess Spurgeon's lecture: as a mockery deserving ridicule.

Spurgeon had been criticized before, but never like this. By delving into the subject of the gorilla, the press said, the evangelist was blurring the lines between theology and theater, between the sanctified and the secular. The fact that he'd invited Layard, the politician, onstage was a sign of a disturbing new trend. According to one reviewer, it seemed that elected officials were now required "to eat dirt and lick the shoes of their constituencies" by pandering to people like Spurgeon, who had a firm grip on the hearts and wallets of his electorate.

"Of course such an exhibition is a disgrace to London," declared an article in the *Literary Budget*. "It is a lamentable disgrace that one of our largest halls can be filled with people ready to listen to a fellow whose sole claim to attention has been his willingness to practise the easiest and most evil form of all buffoonery, that of making jokes over the Bible and burlesquing the inculcation of its truths. One very slight and indirect consequence for good, however, it is possible to detect in these his secular escapades. They have completely vulgarized the practice of amateur lecturing. Now that Spurgeon has taken to lecturing, every young gentleman, who has had a fancy for it, must give it up. He has followed a fashion and destroyed it."

Other reviews were no less devastating. Spurgeon was perplexed by the reaction. He'd been injecting emotion and drama into his sermons for years, but never had he been the subject of so much public scorn. In the weeks following the lecture, he fell ill, but the abuse continued, unabated. Two weeks after the lecture, with Spurgeon too sick to deliver his Sunday sermon, his church's leaders unanimously passed the following resolution:

> *That the members of this church, constantly refreshed*
> *by the gospel ministry of their beloved Pastor, and deeply*
> *obliged to him for the lectures he gives upon secular and*

social subjects, have noticed, with sincere regret, and heart-
felt sympathy with him, the scandals heaped upon his name
by the public press, and beg to express to him their most
loving confidence, their strong desire to endure with him a
full share of his reproach, and their full determination, by
God's help, to bear him constantly on their heart in prayer.

The support buoyed Spurgeon, and a week later he wrote to a friend that the outrage over his gorilla exhibition only strengthened his belief that conceding any ground to the cultural critics would be unforgivable.

"This work of my Institution is of God; lectures are a part of the necessary plan, they do good, I have a call to this work, so all this opposition is a spur to increased zeal," he said.

For decades after this lecture, Spurgeon poured that zeal into an international fight against a relaxed, liberal reading of Scripture. His influence was felt strongly in the United States, where his lectures and sermons were circulated to millions of people and became models for America's new evangelical movement. Preachers like Dwight Moody— whose emotional revivalism helped spearhead the fight against evolutionary theory in the United States—counted Spurgeon a pioneering hero for his determination to battle those who questioned the sanctity of the book of Genesis. Before the evolution debate blurred the lines between men and animals, many people, like Owen, considered a metaphorical reading of the Bible to be compatible with a life of faith. Now an increasing number of evangelicals disavowed any interpretation that strayed from the strictly literal.

"Compromise there can be none," Spurgeon wrote, summing up his combative resistance against those whose liberal readings of the Bible allowed room for heresies such as evolutionary theory. It was a slippery slope, and he wanted no part of it.

"One way or another we must go," he wrote. "Decision is the virtue of the hour. Neither when we have chosen our way can we keep company with those who go the other way."

At Spurgeon's lecture on the gorilla, a new brand of popular evangelicalism fully embraced the battle pitting science against religion, and it would never concede the debate.

Paul seemed uninterested in joining the crusade. While Spurgeon dug in his heels, the adventurer began spending more time among a rowdy crowd who dedicated themselves to destroying the kind of public piety that the evangelist wore on his sleeve.

·∘⟧ ⟦∘·

CHAPTER 28

Mrs. Grundy and the
Cannibal Club

Leicester Square was not a place where tender, pious souls often found themselves after dark. Once fashionable, the neighborhood had grown a little long in the tooth by the 1860s. When the sun went down, the street came to life, crawling with billiard-room sharks who stank of cheap cigars and ladies of ill repute who stank of cheap perfume. Bertolini's, a restaurant in the middle of it all, specialized in both Italian and French food, and the owner—a cheerless old man who called himself either Giovanni Dominico Bertolini or John Dominique Bertolini, depending on which specialty was on the stove—had been treading the dining room floorboards for more than five decades. The food was cheap but tolerable, and the restaurant's grubby charm drew bohemians who liked to slum it every now and then. Tennyson affectionately called it "Dirtolini's."

If, by some devious twist of fate, a devout Victorian stumbled through the closed doors of the banquet room on a Tuesday evening, his fragile sensibilities would have been instantly shattered by the sacrilegious invitations to damnation on display. A small cadre of culture warriors calling themselves the Cannibal Club held court here. They were well-read sophisticates hell-bent on flouting the puritanical rules of decorum they believed strangled the life out of Victorian England. The members of the Cannibal Club celebrated everything that proper English society deemed taboo. Offensive ideas—whether they regarded race, religion, or sex—were aired without apology, in the name of unfettered intellectual inquiry. The only reason the club wasn't the most notorious in England was that few outside the members themselves knew it even existed.

Meetings were called to order when someone banged on a table with a carved wooden mace, which depicted a native African gnawing on a human thighbone. Then, before the group launched into discussions ranging from, say, female circumcision to an anthropological history of erotic flogging, a member would recite the club's mock-religious invocation—a blasphemous send-up of the Eucharist called "The Cannibal Catechism," which depicted the holy Christian sacrament as a cannibalistic ritual.

> *Preserve us from our enemies*
> *Thou who art Lord of suns and skies*
> *Whose meat and drink is flesh in pies*
> *And blood in bowls!*
> *Of thy sweet mercy, damn their eyes*
> *And damn their souls!*

The verses were written by a young Algernon Swinburne, the licentious poet and Cannibal Club mainstay whose professed desire was to make the teakettles of British society "seethe and rage." Other members included Charles Bradlaugh, the first openly atheistic member of Parliament; Henry

Spencer Ashbee and Richard Monckton Milnes, who owned, respectively, the two most extensive stashes of pornography in England; and Richard Francis Burton, the explorer and Renaissance man whose literary translations would eventually introduce much of the English-speaking world to the *Kama Sutra* and the unexpurgated *Arabian Nights*. They'd eat, drink, and let their conversations veer absolutely wherever they wanted. According to legend, Burton (athletic, imposing) occasionally had to carry Swinburne (fragile, pixie-like) out of the restaurant under his arm after the liquor was gone.

The members were drawn to one another thanks to a shared hatred for one "Mrs. Grundy"—a fictional composite who epitomized the tight-laced prudery that threatened to define the era. The name came from a play by Thomas Morton; one of the characters, when worrying about how she'd be viewed by her priggish neighbors, agonized, "What will Mrs. Grundy say?" It became a catchphrase, because England was crawling with Mrs. Grundys. They weren't necessarily women; the Cannibal Club would have considered the Reverend Charles Spurgeon a classic Grundy, especially when he crusaded for temperance. The same went for the energetic members of the Society for the Suppression of Vice, whose public crusades against any literature it deemed morally damaging led to the Obscene Publications Act of 1857. When Burton was working on the *Arabian Nights*, a volume that one reviewer condemned as "absolutely unfit for the Christian population of the nineteenth century," Burton's reaction perfectly summarized the Cannibal Club's defiant brand of moral disobedience: "Mrs. Grundy is already beginning to roar; already I hear the fire of her. And I know her to be an arrant whore, and tell her so, and don't give a goddamn for her."

In the spring and summer of 1861 the core members

of the Cannibal Club first forged their bonds, and Paul was drawn into their circle. The friendships he made would last for years and play a pivotal role in the course of his personal and professional history.

Because Victorian London was a place with ample room for the outwardly pious and the secretly subversive, no acrobatic contortions were necessary for him to maintain one foot on Spurgeon's altar and the other with the most notorious cultural rebels in Britain. All he needed was his talent for assimilation and a well-timed introduction to Burton.

THE FORTY-YEAR-OLD Burton was at a crossroads in 1861. In January he had married Isabel Arundell, a devout and aristocratic Catholic, who promptly pulled strings among her well-placed relatives to secure her husband a position in the government's Foreign Office. If Burton were anyone else, a comfortable transition toward conventional respectability would have been the logical next step. But his utter disregard of convention was already legendary. In the months immediately following his marriage, he seemed doubly determined to declare his independence from tradition.

He'd earned his maverick reputation, in large part, thanks to his undercover journeys to the holy cities of Islam. In 1851 he'd made the hajj to Mecca and Medina disguised as a Persian dervish, and three years later he snuck into the walled city of Harar, in what is now Ethiopia. Both cities were considered off-limits to nonbelievers. His success during those journeys stemmed from much more than a gift for mere surface mimicry. He was dyed to the core with erudition, and he was possessed of a tireless attention to detail. Not only did he speak more than twenty languages; he spoke them well. Little escaped his notice, or his notebook. His uncanny ability—and undisguised eagerness—to deeply penetrate foreign cultures had earned him the nickname "the white nigger" among some of his less-than-admiring contemporaries. Burton himself preferred another nickname: "the amateur barbarian."

He chronicled both of his forbidden journeys in books, which made him famous. They also raised suspicions from those who believed that his charitable descriptions of Islam betrayed an absence of Christian faith. His refusal to be willfully ignorant of anything, no matter its social stigma, fueled rumors about his moral character.

The most damaging story involved his sexuality. As a young officer stationed in what is now Pakistan, Burton was the only member of his unit who could speak Sindhi. As a result, he had been assigned to report on the existence of brothels that were reputedly corrupting British troops. Burton's report was characteristically thorough and unflinchingly explicit, with detailed descriptions of the services available, even those offered by eunuchs and transvestites. The report apparently satisfied his superior, who ordered the brothels destroyed and continued to hold Burton in esteem. But two years later, when the commander was transferred to London, he left Burton's report, or notes taken from it, behind. A new commander found it and was scandalized, apparently assuming that such a wealth of detail could only have been acquired through direct observation or participation. He sent the report to higher-ups in Bombay and recommended that Burton be dismissed from the service. Burton wasn't discharged (he'd only been following orders, after all), but his reputation was smeared. Rumors of sexual deviancy dogged Burton for the rest of his career. Opportunities for promotion consistently passed him by, but his resolution to utter uncomfortable truths only got stronger.

During the 1850s, Burton turned his back on the British military and embraced the Royal Geographical Society. But by early 1861, that relationship was also on shaky footing. The problems stemmed from an expedition he led in 1856 to search for the source of the Nile. Burton's subordinate was John Hanning Speke. Burton's respect for the haughty and highborn Speke, who spoke no African languages, was never strong, but by the end of the expedition in 1858 it had turned into near hatred. Speke had discovered Lake Victoria while Burton was temporarily sidetracked by illness, and he declared that he had discovered the Nile's source, though he lacked evidence to support the conclusion. Burton said that more detailed observations would be necessary before such a claim could be made. The two men agreed to end the expedition and return separately to England, where they would jointly present their results to the RGS.

But Speke arrived home weeks before Burton, and he declared to the group that he had solved the mystery of the Nile. In later years, Speke's reputation would be destroyed by this display of naked ambition and presumption, but, temporarily at least, his power play worked. By the time Burton returned to England, Speke had already secured funding from the RGS for a second expedition to the region to confirm the discovery. He had also bad-mouthed Burton to anyone who would listen, recycling the old suspicions about Burton's moral depravity and suggesting his loyalty to the Crown was dubious. Burton, the son of an English military officer, had spent much of his youth in France and Italy, and Speke believed this diluted his Englishness. When Burton eventually arrived in England and voiced his inevitable displeasure with Speke, the ambitious young subordinate became defensive. Referring to Burton's continental upbringing, Speke told one editor that he would "die a hundred deaths" before he let "a foreigner take from Britain the honour of discovery."

With his reputation freshly assailed, the newly married Burton faced an uncertain future. His best friend in the beginning of 1861 was Monckton Milnes, who introduced him to Swinburne at a party in March (Burton in turn introduced Swinburne to alcohol, according to Milnes, sparking a ruinous love affair with the bottle). By midsummer Burton and the half-dozen core members of the Cannibal Club had become a clique. Weeks later, Burton pulled Paul—who'd already mingled socially with Milnes—into their circle.

PAUL'S BACKGROUND among missionaries certainly wouldn't have impressed Burton, who believed that "Christianizing" native populations generally led to their cultural demise. But Burton had other reasons to be drawn to him. In March the Foreign Office, thanks to his wife's pleading, had finally offered Burton a job: British consul for Fernando Po, a flyblown Spanish island used by the British navy to monitor the slave trade off the coast of West Africa. It was the lowliest consulship in the service, and the unattractive offer reflected the depths to which his reputation had sunk. Burton told Milnes that he accepted "the governmental crumb" only so that one day he might be able to nab a "governmental loaf." Burton's experience in Africa exceeded that of almost anyone in England, but his wanderings had been limited to its eastern and central regions. He had little more than four months—April to August—to learn as much as he could about West Africa before he departed, a period that corresponded exactly with Paul's explosive entry onto the London social scene.

The fact that the young adventurer was an outsider among the learned societies of London would have appealed to Burton, whose recent experiences with the RGS had soured him on "armchair geographers"—those who stayed in England and attended meetings while the real men of action risked their lives. Some of the officers of the RGS barely concealed their contempt for field scientists, viewing them as little more than mindless robots sent out to collect data that could be analyzed by experts back home. Burton hated the idea that explorers were expected "to see and not to think," as he phrased it in the preface to his book about the Nile.

Burton's distaste for this notion was never more intense than in 1861, thanks to his recent arguments with a prominent RGS fellow named W. D. Cooley. Cooley authored a book in 1852 that he triumphantly titled *Inner Africa Laid Open.* He claimed the Dark Continent as his area of expertise, even though he'd never set foot there. After his expeditions, Burton had exposed several damning errors in Cooley's work, and in his usual sardonic style he dismissed one of Cooley's articles as "a most able paper which wanted nothing but the solid basis of accurate data." Cooley responded by publishing a pamphlet stating that Burton's conclusions during the Nile expedition could not be trusted because of an ignorance of the local languages. The pamphlet asserted that Burton had obviously lied about speaking Swahili during his expedition because, according to Cooley, Swahili was unknown in Africa beyond the coast. His charges were supremely confident and entirely wrong.

The attacks against Paul by arrogant skeptics who had rarely ventured out of London seemed similarly unjust to Burton, who unlike Waterton sought to defend Paul from the barbs. In a world of "gentlemen and players," Paul seemed, like Burton, a player.

In May, Paul was invited to speak at a meeting of the Ethnological Society of London by Burton and James Hunt, the group's president. After listening to Paul describe the tribes he'd encountered both on the coast and inland, Burton declared that the descriptions rang true, based on his experiences in eastern and central Africa.

Two months later, Burton prepared a lecture that he titled "Ethnological Notes on M. Du Chaillu's *Explorations and Adventures in Equatorial Africa.*" He invited Paul to hear its presentation at the Ethnological Society.

Burton praised Paul's work, and his only direct criticism of the book was its lack of exactitude in transcribing the native languages—"a sin of omission," Burton said.

"This paper will, I trust, satisfy the most querulous that Monsieur Du Chaillu has well and voraciously studied the new and curious races of whom he has treated," Burton concluded. "For myself, I must be allowed to offer him my best thanks; every page produces upon my mind the effect of the bugle upon the cast charger after a year or two in the cab-shafts of civilization. And I venture to express a hope that at some future day I may be permitted to appear before the Ethnological Society as an eye-witness of, not merely an analogical testimony to, the truthfulness of the picturesque and varied pages which have caused such a sensation on both shores of the Atlantic."

Applause greeted Paul when he stood to say a few words of thanks. But as soon as he began to speak, it was clear that one querulous critic had not been convinced by Burton's show of support. The man began hectoring Paul.

Paul had endured months of ridicule and public abuse. His character had been assailed in some of the most widely read publications in the world. Now, in the middle of a dignified gathering, Paul's composure cracked for the first time. His inner barbarian burst out charging.

THE MAN'S voice kept interrupting Paul with catcalls and taunts.

The voice belonged to Thomas Malone, a thirty-eight-year-old chemist who worked in the laboratory at the London Institution but was better known as an expert in the nascent field of photography. He and a partner had opened one of London's first photographic studios, and he had helped establish the school of photography at the Royal Polytechnic Institution. Malone was also a regular at the meetings of scientific societies, promoting photography as a tool that could be used by almost any man of science, from geographers to ethnologists.

Malone wasn't much interested in Paul's descriptions of gorillas or of the native tribes. He had not read *Explorations and Adventures*, but he had seen an article about it in the *London Review* and knew that Paul had described a stringed musical instrument used by the Fang called the *ombi*. Comparing the instrument to a guitar or a harp, Paul had written that the strings were made of long fibers extracted from the roots of a tree. Malone didn't believe such strings could produce musical sounds and made his doubts clear to everyone in the crowd. When Paul begged to differ, explaining that he had seen and heard the instrument himself, Malone's pestering intensified. He began shouting questions into the air:

*Did you see everything you describe in the book with your
own eyes?*

Did you actually write the book yourself?

Has the public been misled?

"Of course all my remarks were unpalatable to M. Du
Chaillu and his friends, and I was begged to keep myself to
the discussion of Mr. Burton's paper," Malone reported after
the event. "I did, however, tell M. Du Chaillu that the days of
authority with regard to things capable of human proof were
gone, and that any scientific man, however eminent, who came
to us with novel statements must expect to be questioned."

What Malone described as honest inquiry was rendered
as obnoxious browbeating by other witnesses, including Burton and James Hunt. "He then rose," Burton said of Malone,
"and, after a preamble touching the fierceness of his disposition, adopted a tone and style of address which would have
caused the coolest temper to boil over."

Paul stepped away from the podium, humiliated and
enraged. As the audience began to disperse, he searched the
crowd for Malone, stepping over chairs and benches to reach
him. Hunt reported: "Soon after the chairman had left the
meeting, I observed M. du Chaillu making his way in an
excited manner to that part of the room in which Mr. Malone
had been sitting. I followed him, but did not reach the spot
until Mr. Malone called out, 'Is there no one here to protect
me?'"

Several witnesses heard Paul shout at Malone, "Coward!
Coward!"

After the *Globe* newspaper reported one version of events,
Malone attempted to clarify the exact sequence: "I was pre-

paring to go, when, to my astonishment, I saw standing before me, on a form, a little figure with dark threatening eyes and hands. It was M. Du Chaillu. I did not hear or heed what he said, for I instantly received—the outrage described in the *Globe* extract."

Paul's actions had been so *unspeakable* that Malone couldn't even bring himself to spell them out. But in the words of the *Globe*, that outrage consisted of "the wild justice of expectoration." In other words, Paul spit in Malone's face. It was a startling vulgarity that stunned some of the gentlemen present.

Burton didn't count himself among the offended. More propriety was the last thing he believed the Ethnological Society needed. Feeling stifled by the "respectability" of the organization, he and Hunt would soon quit and start a new group, the Anthropological Society of London. The new society, which would become the quasi-official haven for all the members of the Cannibal Club, was intended as a place where Mrs. Grundy would be banished to eternal hellfire.

When the new group was officially inaugurated two years later, Paul was appointed an honorary secretary. The incident with Malone had scandalized the public, but it had won Burton's lifelong friendship.

"My wonder is that M. Du Chaillu had restrained himself for so long," Burton quipped.

But others wondered aloud just who, exactly, Paul was and where he'd learned his manners.

"We fear M. Du Chaillu has been too long among the gorillas," reported one newspaper. "We regret that so distinguished a traveler and author should so disgracefully forget himself, and suggest to an English public whether the gasconading spirit of his Gallic blood has not been intensified and degraded by his naturalization in the land of the bowie-knife and his sojourn in the country of the gorilla. . . . We don't relish these foreign manners; and, if M. Du Chaillu does not wish to appear in the police courts, he had better abandon them, whether they be French, African, or American."

Where, exactly, had Paul come from? He never spoke about his youth, and his silence had, unbeknownst to him, fueled speculation about his background that had already begun to circulate on both sides of the Atlantic.

CHAPTER 29

"Evidence of a Spurious Origin"

Charles Waterton had jealously squared off against rival naturalists before, but no battle shed more light on his ire toward Paul than did his previous effort to discredit John James Audubon.

Waterton had always prided himself as a gifted observer of birds. If he had an ornithological hero, it was Alexander Wilson, a Scotsman whose *American Ornithology* was something of a sacred text for Waterton. That book inspired Waterton to visit North America shortly after *Wanderings in South America* was published, and there he befriended George Ord, a zoologist in Philadelphia who had written a biography of Wilson. Ord and Waterton forged an instant friendship that centered on their reverence for Wilson and their animosity toward Audubon, who was threatening to knock him off his perch as America's foremost bird specialist.

Waterton met Audubon once, in New York in 1826. Compared with the pedigreed English squire, the young French-American bird hunter was a rowdy upstart, a long-haired maverick in a buckskin coat. Audubon delighted in speaking of the untamable nature of the untrammeled American landscape and prized his skill as a rifleman. Sometimes, in his speech and the textual descriptions that accompanied his paintings, Audubon seemed to anthropomorphize his animals, lending them the most indelicate of human traits. When he described the bald eagle, for example, he suggested that the raptor delighted in inflicting suffering on its prey:

> *It is then that you may see the cruel spirit of this dreaded enemy of the feathered race. . . . He presses down his powerful feet and drives his sharp claws deeper than ever into the heart of the swan. He shrieks with delight as he feels the last convulsions of his prey, which has now sunk under his increasing effort to render death as painfully felt as it can possibly be.*

Waterton thought him a charlatan and believed Audubon was falsely elevating animals by ascribing human characteristics to them—a sin he'd live to see repeated by Paul. Together with Ord, Waterton had conducted a transatlantic smear campaign designed to rob Audubon of all scientific credibility. Some of their potshots found their target. They once correctly called Audubon out for claiming to have discovered a new species he called the "Washington sea eagle," when in fact it was nothing more than an immature bald eagle. But they often let their venom cloud their judgment, taking their criticisms too far.

In the 1830s, Audubon painted a mockingbird, placing a rattlesnake with it in a tree. Waterton and Ord insisted the image included at least two mistakes: rattlesnakes cannot climb trees, and their fangs are invariably curved inward, like scythes, and never feature even slight recurvature, as Audubon's did. On both points, Waterton and Ord were wrong.

Rattlesnakes can ascend trees, and sometimes their fangs do curve in the manner Audubon described.

By attacking Audubon's description of a snake, Waterton seemed to be playing with fire. In *Wanderings*, the Squire described fighting off a snake attack by putting his fist into his hat and then stuffing it down the snake's throat, overpowering the ten-foot serpent with raw courage. To some, Waterton had a lot of nerve accusing someone else of taking liberties with the description of a snake in the wild. "Audubon has been rudely assailed about a 'snake story,'" said the Reverend John Bachman, who collaborated with the artist on a book about America's quadrupeds, "but Waterton has given us several stories that fairly fill us with wonder and dismay."

The distaste for Audubon by Waterton and Ord probably had as much to do with class snobbery as scientific disagreement. Audubon was a shadowy figure during his life, and his illegitimate birth in what is now Haiti was a secret he guarded closely. He went so far as to surrender his claims of inheritance from his father to ensure the facts of his birth didn't become public. He had a fake passport that said he was born in New Orleans, and he falsely claimed that he had studied art under the tutelage of Jacques-Louis David, an influential French painter. Audubon was, in some ways, impersonating an educated man of noble birth to gain acceptance in a scientific community that valued such qualities. Ord and Waterton, both highborn sons of wealthy patriarchs, sniffed out inconsistencies and were determined to expose him as one who'd trespassed into spheres that were beyond his station.

Audubon was feted when he visited England in 1826, celebrated by some of the country's leading men of science. Waterton was incensed. "Without leaving behind him in America any public reputation as a naturalist, Mr. Audubon comes to England, and he is immediately pointed out to us an ornithological luminary of the first magnitude," Waterton huffed. "Strange it is, that he, who had been under such a dense cloud of obscurity in his own western latitude, should have broken out so suddenly into such dazzling radiance, the moment he approached our eastern island."

He and Ord redoubled their attack. In a letter, Ord predicted that "many of those who have afforded their patronage to the contemptible imposter, will blush to think that they ever made his acquaintance." For years, they waged a two-man war on Audubon's reputation, which did nothing to prevent him from becoming the most celebrated ornithologist of all time.

In 1861, Waterton must have experienced an unsettling pang of déjà vu. He wasted no time in writing to Ord to tell him that he'd found another trespassing pretender who seemed to be begging for a corrective reproach.

He wrote: "Audubon is immaculate when compared to Du Chaillu."

PAUL WAS everything Audubon had been, only more so: a man who exaggerated the untamable state of nature and who had splashed onto the scene minus the proper pedigree, wooing some of Britain's most celebrated arbiters of culture. Waterton knew that Ord would be very interested in his struggle to expose the so-called American as counterfeit goods.

"I have warmed his hide in the last five or six numbers of the *Gardener's Chronicle*," Waterton informed Ord. "I am quite convinced, in my own mind, that Du Chaillu's adventures in the land of the gorilla are nothing but impudent fables. He always meets the gorilla on the ground. It ought to have been in trees. I suspect strongly that the traveler has been nothing but a trader on the western coast of Africa; possibly in kidnapping negroes, and that he has bought his skins of negromerchants from the interior."

Gossiping about an unfounded connection between Paul and the illegal slave trade must have seemed a juicy tidbit thrown in for Ord's pleasure. But as Ord began asking his friends within the Philadelphia academy about Paul, he stumbled across an explosive notion: *What if this Great White Hunter wasn't really white?*

The rumor had spread around Philadelphia that Paul might be half-black. Ord suggested that this secret might explain his soured relations with his onetime sponsors in Philadelphia. Some of the members of the academy, Ord told Waterton, had taken note of "the conformation of his head, and his features" and had detected "evidence of a spurious origin."

"If it be a fact that he is a mongrel, or a mustee, as the mixed race are termed in the West Indies," Ord wrote to Waterton, "then we may account for his wondrous narratives; for I have observed that it is characteristic of the negro race, and their admixtures, to be affected by habits of romance."

It was a twist Paul probably saw coming, but he was powerless to stop it. In the forests of Africa, he had taken control of his destiny, slain a beast no one had ever seen, and effectively recast his life as a heroic myth. But his past reared up to consume the identity he had risked his life to create. It proved a much harder beast to kill than a gorilla.

·◌⫴◌·

CHAPTER 30

Shadows of the Past

Paul's early history was an involute knot that he had deliberately tied himself. Throughout his life, he spoke little of his French father, Charles-Alexis Du Chaillu, the trader on the coast of Gabon, and never publicly spoke of his mother. Sometimes he'd hint that there were Huguenots (persecuted French Protestants who'd been forced to leave France) among his forebears. Other times he'd vaguely refer to a "Creole" ancestry. That word conjured images of steamy bayous in the Mississippi delta, and when people assumed that Paul was from New Orleans, he didn't correct them. After his father died in the mid-1850s, he began using the name Paul Belloni Du Chaillu. People naturally assumed that "Belloni," with its Italian bounce, came from his mother's side of the family. It made sense. A southern Mediterranean origin might explain the dark eyes and olive skin.

If a person had asked Paul's friends in both America and England where they believed he had been born, the variety of answers would have exposed Paul's falsifications. Many would have said his birthplace was Paris. Those answering America would have disagreed about the state: some would have sworn he'd been raised in Louisiana, some would have picked New York, others would have said South Carolina.

But of all the conflicting origin stories, one stood out. As far as can be determined through the historical record, he told it to only one person.

In London, Paul befriended a man named Edward Clodd, an eminent banker who occasionally wrote articles for the scientific and literary press. Clodd welcomed Paul into his social circle, which included some of the most prominent literary men in England. Paul also visited the Clodd home regularly, dining with the wife and entertaining the children with his stories of his adventures.

At some point, Paul confessed to Clodd that he was neither French nor American. He said that he had actually been born on a French-controlled island off the coast of Africa. Paul's father, it was true, had been born in France. But his mother was a native.

Clodd recognized that the information could have ruined Paul's reputation. He guarded the secret.

THE ISLAND of Réunion (until 1793 known as Bourbon) sits in the Indian Ocean, off the coast of Africa, east of Madagascar. The French colonized the previously uninhabited territory in the late seventeenth century. They brought with them slaves from Madagascar and the African mainland. Throughout the eighteenth century, the population of the island was roughly half white, half black. But in the late eighteenth century, the slave trade boomed, and blacks became the overwhelming majority. By the century's end, 61,300 people lived on the island, and about 85 percent were black. Most were slaves, but about 1,200 were known as "free domestics" or "free coloreds" who worked for the white landowners without being considered their property. The two cultures—French and African—increasingly blended throughout the nineteenth century, and the island became known as a racial melting pot. A pronounced Creole culture—not white, not black, but unique unto itself—took root.

Unlike in the United States, children born of a white man and a black woman in Réunion generally became part of the man's family. This allowed many of the island's mixed-race children to escape the bonds of slavery, though they often occupied a lowly status in the father's home.

Existing archival records offer patchy glimpses into Paul's likely childhood. The evidence points to a strained relationship between a mercurial father and an overlooked son for whom "home" would always be a shifting concept.

IN 1831, the year of Paul's birth, Charles-Alexis Du Chaillu was earning a reputation as a dangerous man. Ever since he'd moved to Réunion from eastern France, the thirty-year-old trader had struggled to climb the island's social ladder. Even though he was among the island's white minority, he could only rise so far because he was not one of the plantation owners known locally as the sugar aristocracy. They made the rules on the island, making sure everyone else was kept in his proper place.

Around 1830, Charles-Alexis and a group of like-minded white middle-class businessmen supported a proposal to create an elected assembly of representatives. The plantation owners hated the idea. They had always been able to exert their influence on the island's appointed governor, and they feared any changes to the status quo. They thought the upstarts clamoring for an assembly were troublemakers who seemed to be growing braver with each passing day.

In the summer of 1830, King Charles X of France was overthrown in a coup called the July Revolution. The governor of Réunion, who protected the interests of the elite sugar growers, was a loyal appointee of the king. Charles-Alexis saw the death of King Charles X as an opportunity that might allow him to topple the local government.

The governor knew he was vulnerable, and he tried to preserve the validity of his rule. He confiscated newspapers and incoming mail deliveries, keeping all of the island's inhabitants in the dark about events in Europe. But, inevitably, one passenger arriving by ship snuck by the dockside inspectors with a copy of the *Journal du Havre* in his luggage, and Charles-Alexis began spreading the news. A short time later, a French ship docked in the port of Saint-Denis bearing a new flag with red, white, and blue stripes. The familiar fleur-de-lis banner of King Charles X was gone, replaced with the tricolor symbol of the new king, Louis Philippe. For Charles-Alexis it symbolized a new beginning.

He ran to the port to see it, along with hundreds of others. Someone suggested hoisting the new flag on top of a local naval ship—a dramatic gesture, full of explosive symbolism. Police at the dock tried to squash the plan. But Charles-Alexis and three other men couldn't be stopped.

Holding the flag, they jumped aboard the naval ship, and Charles-Alexis began to climb the tall mast of the ship, where he ripped down the fleur-de-lis and affixed the banner of King Louis Philippe. Shouts of "Vive la France!" and "Long live freedom!" rang over the water. High above the port, looking down upon the cheering crowd, Charles-Alexis might have believed that he had finally scraped his way to the top of

Réunion's social scale. But it was a brief apogee. High on the mast, he was a sitting duck—an easily identifiable enemy of a political establishment that wasn't ready to give up.

For the next several months, tensions ran high between the island's conservative elite and the supporters of King Louis Philippe. On May 1, 1831—exactly three months before Paul's birth—a group aligned with the sugar barons clashed with the reformers. Around the same time, a ship registered to Charles-Alexis was detained by authorities. Unregistered blacks—perhaps slaves—were found on board. The governor, who still hadn't been replaced by the new king, saw an opportunity to rid himself of the most persistent provocateur on the island. He exiled Charles-Alexis from Saint-Denis for one year and subjected him to police surveillance. Charles-Alexis chose to be deported to France.

Government records reveal that he sailed to Nantes, in western France, in the summer of 1831. This indicates that he left the island before his son was born on July 31. For several years after this, Charles-Alexis was absent from the island's archives. But he reappeared in the decadal census of 1840.

Between his exile and his return, Charles-Alexis married a woman named Marie-Julie Bréon, who was born in Saint-Denis but was not Paul's mother. By 1843, their household was thriving. Two children—both daughters—were registered with the census authorities, along with seven slaves. Paul wasn't recognized as an official member of the family.

Although island tradition would put Paul in his father's custody, the exile of Charles-Alexis would have prevented that arrangement during the first years of Paul's life. However, it's possible that he joined Charles-Alexis and Bréon after his father's return, even though his name doesn't appear on the official registry. Henry Bucher, a historian who in the

1970s uncovered many of the records pertaining to the family, reported that it was not uncommon for illegitimate children to remain unregistered, even though they lived in the household.

The woman who gave birth to Paul remains shrouded in mystery, but some clues exist. A Gabonese historian, Annie Merlet, in 2007 consulted the Cultural Services Department in Saint-Denis and confirmed that "Belloni" was among the names used by the "free coloreds" in Réunion. She also discovered that Charles-Alexis had been accused of "immorality" resulting from dalliances with women of black or mixed-race ancestry. Merlet speculated that Paul might have been raised by his mother until her death in the 1840s, at which time the boy sought out his father, who had since moved on to Gabon to manage the Maison Lamoisse of Le Havre trading company.

Paul first arrived in Gabon in 1848, but he might have spent some time in Paris between his years in Réunion and those in Gabon. A student of his in Carmel, New York, recalled that Paul spoke of the 1848 French Revolution, which deposed King Louis Philippe, as if he'd been there himself. The student remembered that Paul clearly held Louis Philippe in high esteem and was horrified by his violent overthrow. "He hated the country in which such things could be," she said, believing that it was a principal reason he had so enthusiastically adopted America as a homeland. It's impossible to know whether or not Paul was in France in 1848, but there's no reason to believe he wasn't. His loyalty to the deposed king would make perfect sense for a boy raised on Réunion, where Louis Philippe had been revered by his father's generation of reformers.

When Paul arrived in Gabon as a teenager, Charles-Alexis

was in charge of all material provisions for the French colony and its naval installations in Gabon. In 1850, Gabon's administrative authority listed Paul as "a very young" assistant in the local trading storehouse, which suggests he might have been officially employed for a time by his father. But Paul's eager embrace of the Wilsons as a surrogate family implies that the father-son relationship was untraditional and distant and likely strained.

After he traveled to the United States in 1852, Paul was able to maintain the fiction that he'd been born an American among many of his acquaintances because passports weren't necessary for international travel. According to Putnam County, New York, records, he did, in fact, apply to become a naturalized American citizen in 1855. But his request wasn't honored. No reasons for the apparent rejection are known, but proof of parentage—both paternal and maternal—was required of all applicants.

WHILE THE controversy about Paul's credentials as a naturalist and explorer raged, the rumors of his mixed-race ancestry circulated behind closed doors. The gossip likely eroded his standing in Philadelphia and might have fueled some of Charles Waterton's diatribes, but the general public that followed the controversy in the press remained unaware of those covert machinations.

It simply wouldn't have occurred to most people who had met him that Paul was anything other than European. In the pictures that survive of him, his prematurely thinning hair, though usually trimmed short, appears more straight than curly. Paul himself made references to his "very dark brown" or "almost black" skin tone in *Explorations and Adventures*, but he always attributed it to excessive exposure to the sun, not to heredity. Journalists occasionally referred to him as "dark" or "swarthy" in print, but they never directly questioned his lineage. "M. du Chaillu is a bald, bronzed, diminutive shrunken specimen of humanity who looks as if Equatorial Africa had absorbed his lifeblood," one newspaper wrote of him. Another journalist, the London correspondent for the *New York Times*, in the summer of 1861 tacked the following to the end of an article about Paul as a provocative afterthought: "By the way, it seems to be another disputed point as to what his politics are. He is an American citizen, and the suspicion of Negro sympathies hangs round him in many ways."

The threat of exposure was never far away, and sometimes hovered closer than Paul could have realized. Years later, Mary Kingsley, who in the 1890s became a pioneering female explorer of western Africa, revealed in an unpublished letter to Clodd that she once possessed the manuscript of Paul's biography "written by an old enemy of his and sent to me for publication which would have blown the roof off any publisher's house in London—not that I have shown it to any one." Kingsley died in 1900. The biography never surfaced.

If it had been revealed that his mother was half-black, Paul would have been called a quadroon. The label would have precluded his acceptance into the highest professional circles in Victorian London, and the connections that fueled his meteoric rise to fame—Owen, Murchison, Murray, and others—would have evaporated. Those who supported him were members of an elite that was infamously intolerant of non-whites. Even some of the most liberal thinkers who adamantly opposed slavery—a list that included Huxley—believed Africans and aboriginal Australians represented the lower stages of human development. Racial minorities didn't belong to the Royal Geographical Society. Their exclusion wasn't debated. It was taken for granted.

The title character in *The Octoroon*—the play that was staged on Broadway while Paul had been in New York—despairs that a single drop of African blood makes her "an unclean thing." Paul would have been similarly stigmatized among many in a population obsessed with whiteness. Some ladies in Victorian England actually ate pure clay, believing that it would lighten their complexions, and others rubbed highly toxic arsenic "complexion wafers" over their skin for the same effect. After visiting England twice, Ralph Waldo Emerson in 1856 noted a marked tendency to overvalue racial status among the British of that era. "Men hear gladly of the power of blood or race," he wrote in a book he titled *English Traits*. "Every body likes to know that his advantages cannot be attributed to air, soil, sea, or to local wealth, as mines and quarries, nor to laws and traditions, nor to fortune; but to superior brain, as it makes the praise more personal to him."

Paul, unintentionally, was helping to make that trait even more pronounced with his descriptions of the native Africans who shared the same forest with the gorilla. Even as his secret remained successfully hidden from most people, Paul found himself walking through a new world where a person's ancestry seemed to matter more than it ever had.

CHAPTER 31

Black and White

Compared with most other explorers of his era, Paul was considerably kinder in his portrayal of native Africans in *Explorations and Adventures*. Still, at times he adopted the paternalistic tone that was de rigueur among Victorian exploration narratives—the white conqueror strolling among hapless savages—but at other times he'd slide into the lowly role of the subject. On one page he defined the African character as essentially deceitful, but then a few pages later he launched into praise for the "kind-hearted negroes" he traveled among.

He wrote: "I met every where in my travels men and women honest, well-meaning, and in every way entitled to respect and trust; and the very fact that a white man could travel alone, single-handed and without powerful backers, through this rude country without being molested or robbed, is sufficient evidence that the negro race is not unkindly natured." When among the Mbondemo tribe, he reported that in becoming known as "Mbene's white man," he surrendered to the authority of the chief. "The title has comfort and safety in it," Paul wrote. He had assumed the role of the king's subject, not his overlord. It's difficult to imagine other celebrated

African explorers of the period, such as Burton, Stanley, and Speke, making such a concession. Paul's depictions were certainly not free from condescension, but they were significantly less offensive than others of the period.

His complex perspective is apparent in his portrayal of the Fang tribe. Every lurid detail he included—from the Fang's razor-sharp teeth, to their garish war paint, to his fears that they might "be seized of a passionate desire to taste of me"—supported the idea that the deeper one ventured into the dark heart of Africa, the more savage the natives. But Paul flatly contradicted the commonly held assumption that a tribe's degree of savagery corresponded to its degree of "blackness." He wrote that the Fang were not only the most terrifying cannibals imaginable but also *whiter* than any of the other tribes he encountered.

Those who supported the dawning theory of evolution often assumed that blacks were "less evolved" than whites. Advances in biology and genetics have since obliterated that notion, but in the 1860s some of the most celebrated scientists of the era worked under that assumption. This belief implied that blacks were more closely related to apes than whites were. At times in the narrative, Paul seemed eager to undermine the idea. When he wrote of killing a chimpanzee, he reported that the native hunters took note of the animal's light facial skin. "They roar with laughter," he writes in describing the natives' reaction to inspecting the chimp's seemingly Caucasian appearance. "Look! He got straight hair, all same as you. See white face of your cousin from the bush! He is nearer to you than a gorilla is to us!"

Paul's unique perspective on racial subjects and his relatively sympathetic portraits of Africans did nothing to silence the bigotry seething inside many of his readers. The animal he had shown the world was already becoming a charged symbol of the very hatred from which he hid.

ONE WEEK after Paul first addressed the Royal Geographical Society in London, Abraham Lincoln was inaugurated president of the United States. By late summer, the Civil War was in full swing. *Harper's Weekly*, a popular magazine owned by the same company that published the American edition of Paul's book, began inserting articles about the English gorilla craze in between its war reports. Paul's book became an instant hit in the thirty-four states.

History remembers Edwin Stanton, Abraham Lincoln's secretary of war, for uttering one of the most eloquent epitaphs of all time after Lincoln's assassination: "Now he belongs to the ages." But four years earlier, when he was still an outspoken critic of the president, Stanton nicknamed Lincoln "the Original Gorilla." In a potshot he enjoyed repeating, Stanton added, "Du Chaillu was a fool to wander all the way to Africa in search of what he could so easily have found at Springfield, Illinois."

The nickname stuck, particularly in the newly formed Confederacy. "Since the Southerners have adopted the habit of calling Mr. Lincoln the 'Gorilla,' we observe that some of the papers are giving detailed sketches of that animal," read an editorial in the *White Cloud Kansas Chief* in August 1861. "So traitor blackguardism is doing some good—it is making the people acquainted with the history of a remarkable species of the animal kingdom."

Paul and the gorilla craze he inspired unintentionally gave the less principled ideological combatants in both the Union and the Confederacy a wealth of material to twist to their own ends. Throughout the United States, the term "gorilla" became the ultimate dehumanizing epithet for blacks. When two separate sexual affairs between black men and white women were reported in Detroit (scandals that merited national coverage in 1861), the *Cleveland Plain Dealer* referenced Paul's book to slander all parties involved. "We do believe that if the African Gorilla was plenty in our midst, 'lying around loose' like many of these worthless Negroes, they would have no trouble in contracting 'advantageous family alliances' among some of our women!" When the issue of official emancipation was first debated in Congress, Representative John Law of Indiana predicted that ending slavery would effectively empower "these human gorillas to murder their masters" and rape their wives and daughters.

The gorilla was quickly enlisted to justify the practice of slavery. T. W. MacMahon, a popular essayist from Virginia, in 1862 wrote a book called *Cause and Contrast*, which sold more than five thousand copies in its first week on sale. The author's principal aim was to systematically prove that blacks were saddled with a "hopelessly degraded intellectual organization" and that whites therefore need not apologize for treating them as inferiors. MacMahon called upon a host of "scientific facts" from dubious sources, including the testimony of a Dutch doctor who asserted, "The pelvis of the male negro in the strength and density of its substance, and of the bones which compose it, resembles the pelvis of a wild beast." MacMahon also marshaled Paul to his cause, noting the "awful likeness" that the adventurer perceived between the gorilla and humans. MacMahon apparently believed the observation was full of significance—provided that the humans in question were black ones, like the Hottentots of South Africa, whom MacMahon considered especially beastly. "The Negro proper is certainly not so low in the scale of physical organism as the gorilla," MacMahon conceded, "yet it is demonstrable that he (especially the Hottentot), most certainly approximates in the structure of his frame to the monkey kind and the troglodyte."

From there, it got worse. The Confederate writer Marvin T. Wheat published a 595-page book arguing that blacks should be classified under the biblical category of "living creatures" or "beasts of the earth"—not humans. Genesis stated that God had created man "in His own likeness," and according to Wheat blacks didn't qualify for inclusion in that category, because God's likeness couldn't be both black *and* white. In one perfectly representative sentence, Wheat unrolls a rope of fraying logic:

> *In this there seems to be a palpable contradiction, for it is irreconcilable with natural philosophy, to suppose for a moment, that the two colors, distinct in their natures and organizations, could be created after the Image of One Being, for this being must have had color, as well as other natural characteristics, or he was not nor is a being; and hence we would infer that, speaking technically, philosophically and phrenologically, there could have been but one race of man created after the Image of the Creator, and that all others were created subordinate to him, filling intermediate positions between him and the lower scale of animated nature.*

Based on such articles of "organic law," Wheat declared that slavery was therefore a "Divine Institution" that God commanded white men to follow. The validity of these truths was confirmed by African explorers, Wheat explained, since they exposed blacks as hopelessly primitive. By opening Africa to American and European expansion, Wheat argued, explorers were fulfilling the white man's duty of absolute dominion.

If P. T. Barnum were ever forced to defend the idea that "there's a sucker born every minute," Wheat might have been a convincing Exhibit A. Wheat genuinely believed that Barnum's "What Is It?"—the black man with microcephaly—was a new and undiscovered species of primate. Wheat wrote that the "What Is It?" was, perhaps, a link that connected gorillas with blacks, who deserved to be grouped with apes.

It wasn't Paul's fault that his work influenced such writings, because the tortured reasoning of the writers could have twisted almost anything to serve their ends. But by the second half of 1861 he was clearly uncomfortable with the idea that he'd unintentionally helped blur the lines between the human and the animal worlds. Increasingly, he made a point to tell audiences that he believed gorillas were fundamentally different from humans of all races, echoing Owen's conclusions. Anyone who exploited the supposed links between gorillas and humans to make arguments related to evolution or to racial politics, Paul suggested, was seriously misguided.

AT THE same time that Paul's notoriety reached its height in England, a young Welshman named John Rowlands was struggling to create a new life for himself in the United States. Rowlands began telling people he'd been born in Louisiana, even though he'd been an illegitimate child who'd spent much of his youth in a workhouse for poor orphans. Shortly before the Civil War began, he changed his name to Henry Morton Stanley.

Two decades later, Stanley would assume the role in the public imagination that Paul currently held—Africa's greatest explorer. For both men, the desire to explore and dive headlong into adventure was intricately bound to a desire to transcend the circumstances of birth. Stanley's biographer Tim Jeal has written that for Stanley adventure was "a Nietzschean confrontation that was the breath of life to him, a breaking away from the daily self he knew and could not endure, into a persona in which he could escape past humiliations, and stretch the boundaries of the human condition in denial of his own mortality." Stanley, in his own autobiography, supported that theory when he wrote that he'd been able to find "independence of mind" while exploring Africa. Only there, Stanley wrote, could he achieve a transformative state of presence that "is not repressed by fear, nor depressed by ridicule and insults . . . but now preens itself, and soars free and unrestrained; which liberty, to a vivid mind, imperceptibly changes the whole man."

Stanley's motivating secrets, however, weren't nearly as threatening as Paul's. Being illegitimate was one thing, but being illegitimate *and* part black was incalculably more serious in the eyes of the Victorian establishment. The mania for Caucasian "purity" often meant that people of mixed-race ancestry languished even lower on the social scale than blacks. Richard F. Burton tapped into this sentiment when he wrote about mulattoes as being "neither fish nor fowl" and "despised by the progenitors of both races."

Stanley himself demonstrated the malign potential of Paul's ancestry. On the island of Zanzibar, which like Réunion had become a melting pot of black and white cultures, Stanley lashed out against the progeny of this commingling.

"For the half-castes I have great contempt," Stanley wrote. "They are neither black nor white, neither good nor bad, neither to be admired nor hated. . . . Cringing and hypocritical, cowardly and debased, treacherous and mean, I have always found [the half-caste]. . . . When he swears most, you may be sure he lies most, and yet this is the breed which is multiplied most at Zanzibar."

Such were the views that African exploration routinely excited in England and America and that Paul endured in silence.

Years later, however, with his ancestry still hidden from the public, the *Cincinnati Daily Enquirer* reported that Paul had directly spoken of the mixing of races during a public appearance. According to the paper, Paul believed that it "degraded the superior race without permanently elevating the inferior race." The article ended, "M. du Chaillu is, evidently, not much of a believer in Negro equality."

CHAPTER 32

The Impostors

The press, once so friendly, was ripping Paul's reputation to shreds. He needed character witnesses, and not just from new acquaintances in London like Burton. His best supporters would be those who knew him in Gabon and who could vouch for his credibility. He wrote to the missionaries requesting they write letters in his defense to the London newspapers. Paul promised Owen and his friends at the Royal Geographical Society, the men who'd backed him from the start, that "the truth would right itself in the end."

Months passed. But no letters arrived.

Wilson, at the very least, should have responded. Paul had written to the Presbyterian mission office in New York, assuming it could forward his request to Wilson in South Carolina. But with the Civil War under way, regular mail service between the North and the South had been suspended. So while Paul anxiously awaited the support of a man who probably knew as much about Gabon as anyone else alive, Wilson went about his life in the war-torn Confederacy, unaware that Paul had told his friends and members of the press in London that Wilson's corroboration would prove that his journeys weren't inventions.

In the early fall, Paul sought refuge at the country estates of high-placed friends. During one of these escapes, he got his hands on a copy of the *Athenaeum*, where he saw his name mentioned in a column called "Our Weekly Gossip." The piece reminded readers that Paul had insisted that the missionaries in Gabon would defend his character and confirm his honesty. The newspaper item announced that now rumors were circulating that at long last the first of those responses had trickled into London.

"Letters came to town from the Gaboon on Wednesday last," the newspaper reported. "A gentleman, who has been conspicuously named as a witness for M. Du Chaillu, has written on the subject to his friend in London:—will M. Du Chaillu produce the letter so sent? We hope he will."

Finally, he thought. Paul figured that if the letter had arrived from Africa from someone he had "conspicuously named as a witness," it must have been sent by the Reverend William Walker, who'd taken over as head of the Baraka mission when John Leighton Wilson left the country. Paul had known Walker since 1848, and he'd stayed with him at the mission several times between his inland journeys. Paul had always been friendly with Walker, who would surely stand up for him.

But the *Athenaeum* item was puzzling. Paul should have been notified if the letter had been addressed to him, but he hadn't heard a word about it. Was he the "friend in London," or was the letter headed to someone else?

"Everything will be right," Paul wrote to his publisher John Murray the day after he saw the item in the paper. "The imposters will be those who have tried to blacken my private character and who hate me without a cause."

The next day, a package of letters from Murray dashed all hopes of tranquillity. The letter from Gabon hadn't been addressed to Paul. In fact, it had just been printed in the *Morning Advertiser* newspaper. Murray informed Paul that the letter had been signed by Walker, but it was full of devastating accusations.

Paul couldn't believe what he was reading. According to Murray, Walker's letter ridiculed Paul's knowledge of native languages, and Walker flatly stated that he had invented his stories. "I, in common with most persons, doubt that M. Du Chaillu ever killed or assisted to kill a Gorilla," Walker had written, "and also of the extent of his 'travels.'"

Paul was stunned. How could Reverend Walker have accused him of something like that? It contradicted everything he believed he knew about the man. It didn't make any sense at all.

Unless the letter wasn't really from Walker.

THE REALIZATION hit him like a shot: he had met *another* Walker in Gabon who had nothing to do with the missionary.

He was an English-born trader based in the far south, and his name was Robert Bruce Napoleon Walker—better known in Gabon as either R. B. N. Walker or just Brucie. He was about the same age as Paul. He'd moved from Sussex to Gabon in 1851. His older brother was operating a small "trading station" from aboard a ship docked off the coast, about two hundred miles south of Baraka. In 1857, while Paul was exploring the interior, Walker opened a small station on land near the coastal village of Sette Cama, where he served as an agent for a Liverpool trading firm.

When Paul traveled to the Fernan-Vaz region to establish a base for his southern foray into the jungle, he'd met

Walker several times. He'd told him about his expeditions, his encounters with gorillas, his travels into unexplored territories, and his interactions with native tribes. When Paul let it slip that he hoped to someday visit England and display some of his natural history specimens there, Walker offered to help. He gave him the names of a few friends and relatives in London who might be able to assist him. Walker was a friendly man—or so Paul had believed.

As soon as he figured out that the Walker referred to in the *Athenaeum* wasn't the missionary, Paul dashed off a letter to explain the confusion to Murray. *Still* it didn't add up. Why would R. B. N. Walker want to smear his name? His only interactions with the man had been friendly ones.

"My enemies will not let me rest," he wrote to Murray, asking for a full copy of Walker's letter. "I'm astonished that Walker has written anything against me."

A day later, Paul's copy of the *Morning Advertiser* arrived. It couldn't have been worse. Every sentence stabbed at the heart of Paul's credibility.

"Having known M. Du Chaillu for some years personally, and possessing, moreover, from reliable sources, information the most exact as to his antecedents, besides having a knowledge of many of the places and people which he pretends to describe, I am induced to request a place in your journal for the following remarks."

Walker didn't get into specifics concerning Paul's "antecedents," thankfully, but he implied that the young adventurer was hiding something scandalous about his true identity. Walker suggested that members of a certain trading firm in Paris could provide verification of highly sensitive personal details if informed that the young man claiming the surname "Du Chaillu" was actually "identical with M. Paul Belloni."

The letter spread throughout London like a virus. The *Athenaeum* reprinted Walker's letter the next week, although the editors cut the insinuating reference to his maternal name "for the sake of greater courtesy to M. Du Chaillu." The act suggested that the editors knew very well that Paul's racial background was a matter of rumor and that merely hinting at such a secret was too devastating a claim to risk without irrefutable proof. The *Times*, for its part, declined to print even an excerpt of Walker's letter. John Gray didn't exercise the same restraint. He reprinted the letter, complete with the doubts about Paul's "antecedents," in his monthly journal, the *Annals of Natural History.*

The letter cast doubt on almost every story Paul had told. Walker wrote that the descriptions of the Fang cannibals were greatly exaggerated, if not wholly invented. He asserted that baby gorillas were not "untamable" and added that he'd kept one alive on the coast for several months. He mocked Paul's knowledge of native languages as being "of the most infinitesimal kind." He charged that the so-called gorilla hunter had prepared some of his skins on the coast, not in the jungle, and that Paul had downplayed the presence of other white men in Gabon to make his own limited travels seem more exceptional.

"I think I have sufficiently shown that M. Du Chaillu has been guilty of many incorrect statements; in fact, his work contains nearly as many errors and inaccuracies as there are paragraphs," Walker concluded.

WHAT HAD happened to the friendly R. B. N. Walker whom Paul had met on the Gabonese coast? Had that man been a mere hallucination?

Details emerged that helped explain the betrayal.

After R. B. N. Walker had met Paul in Gabon, he had written to a brother-in-law in England in 1858 and 1859. In both letters, Walker had praised Paul as an admirable young man.

On November 4, 1858, Walker wrote that Paul hoped to travel to England with the specimens he'd collected in Gabon, and he encouraged his brother-in-law to help him make contacts with the professional societies there. "An enterprising naturalist with whom I am acquainted, Mr. Paul Duchaillu (a Frenchman), will shortly be in England with a collection of rare birds and animals, many discovered by him," Walker had written. "I will endeavor to get him to call on you. He has traveled where no white man ever penetrated before." Six months after that initial letter, Walker wrote again, going even further in his praise of the adventurer. He compared Paul to Nimrod, a great-grandson of the biblical Noah and a great hunter who, according to some interpretations of the Old Testament popular in the nineteenth century, was a progenitor of the black race. Walker wrote:

> Mr. Paul Duchaillu, the West African Nimrod, will shortly leave the coast for the United States, and thence to London. I shall give him a letter of introduction to you, and have advised him to get you to revise his journal previous to publication. I shall consider it a favour if you will put him in the way of finding the best market for his specimens, as he deserves to be well remunerated for his trouble. I presume he is about the only European who has seen the N'jena, or Gorilla, in its wild state, and killed it himself. . . . As Mr. D. will therefore be a celebrity in a small way, it will be a feather in your cap to be his cicerone, and to lionize him. I may give

him a line or two to one or two other people; but I think you
are the most likely to be useful to him. He is a very nice little
fellow, and he will amuse you with his descriptions of tribes
and people, who never yet have been seen by another white
man than himself, and who have seen no other than him. As
you will see him in a few months, I will not forestall him by
recording any of his adventures here. He is no boaster, and I,
for one, place confidence in all he has told me, and I consider
that what he related may be relied on.

At first Paul suspected that jealousy might have been behind Walker's stark reversal.

"I cannot express to you the sorrow I felt when I read the letter of one whom I considered as one of my friends formerly, one who gave me letters of introduction to his family, one who begged his friend to present me to English society, one who wrote in one of his letters a most flattering account of me," Paul wrote to Murray as soon as he saw Walker's letter in the *Morning Advertiser.* "Probably he is vexed that I did not mention his name in my book."

Soon, another explanation came to light: Walker himself hoped to cash in on England's gorilla fever.

Within months, R. B. N. Walker would board a steamer for Liverpool with the partially preserved specimens of several gorillas. He had arranged to exhibit the specimens in a museum, and he had organized his own lecture tour.

For the rest of his life, Walker presented himself not simply as a trader but as an African explorer. He supplied museums in England with zoological specimens throughout the 1860s and 1870s, became a fellow of the Royal Geographical Society, attempted ill-fated exploratory expeditions of the Ogowé River, and spoke often of his dreams of writing a book about West Africa. He seemed to want nothing more than to transform himself into another Paul Du Chaillu.

Walker never entirely dropped his attempt to undermine the reputation of the man whose path he sought to follow.

"I am going to pen a few lines for the *Athenaeum* in contradiction of the alleged parentage of Du Chaillu," Walker wrote in a letter to the curator of the Liverpool Museum after he had established himself in London. But if he wrote such an exposé, it was never published. It has been speculated, though never proved, that Walker's biographical sketch was the work that Mary Kingsley later claimed would have "blown the roof off any publisher's house in London" had it been published.

Incredibly, Walker enthusiastically entertained the notion of collaborating with Paul on a joint Ogowé River expedition, and he tried to get the RGS to sponsor such a venture. That collaboration, not surprisingly, never happened.

··◦[ ]◦··

CHAPTER 33

Shortcuts to Glory

The public fallout from Walker's letter was disastrous. Newspapers all over the world publicly accused Paul of inventing everything in his book—the route of his travels, his encounters with gorillas, and even his own identity.

"Belloni, it appears, is the traveler's proper name, not Chaillu, and it would be satisfactory to know the reason of his assuming the alias, which in itself is suspicious," reported the *Glasgow Examiner*. "Until M. Du Chaillu refutes the evidence he himself has invoked, it will be difficult for his best friends to clear him from the imputation of having attempted to pass off his fictitious adventures as a true account of travels that had never been accomplished."

Paul's first public appearance after Walker's letter appeared was in Glasgow. He marched onto the lecture stage inside the city's municipal building. With two stuffed gorillas flanking the podium, he insisted that he had visited every location he had described. Anyone who had the nerve to suggest he had not faced down the frightening beast that stood beside him onstage, he said, should have the courage of his convictions.

"I simply told what I saw," he told the crowd, "and if

any one does not believe me, they had better go and see for themselves."

The audience laughed, unaware that one enterprising author in London was preparing to do just that.

WINWOOD READE had published three books before he turned twenty-three—an impressive accomplishment if all three hadn't unified the nation's literary critics in a chorus of unrestrained contempt. His first novel, *Charlotte and Myra*, was dismissed by the *Athenaeum* as "foolish." His second, a university novel called *Liberty Hall, Oxon*, was skewered by the *British Quarterly Review* as "one of the most untruthful narratives, while affecting to be truthful, we have ever read." The *Spectator* called that book "nauseous" and deemed it "full of all sorts of things he would have done well to erase the moment they were written." In early 1861, Reade published *The Veil of Isis*, a book "deformed by not a little bad taste," according to the *Spectator.*

Now Reade was facing a crisis of professional desperation. He was young, free from obligations, awash in family money—but bereft of literary inspiration. As the controversy surrounding Paul unfolded, however, he sniffed an opportunity. As Reade himself later phrased it, he believed he had "discovered a short cut to glory" in *l'affaire du Chaillu.*

The controversy, Reade saw, needed a referee. Someone had to sort out truth from fiction. Reade appointed himself to the job.

In the autumn of 1861 he began planning a trip to Gabon. There, he would retrace Paul's steps, hunt gorillas, and deliver the definitive verdict on both the adventurer and his descriptions of his infamous beasts.

"In my humble character as a mere collector of evidence," Reade explained, "having no special views to promote, I wish only to arrive at the truth."

ON CHRISTMAS EVE, Reade boarded a steamer in Liverpool. Fifty-one days later, he stepped onto Gabonese soil.

With the five men he had hired during a brief stop in Liberia ("all of whom were tolerably accomplished thieves"), he spent his first night in Gabon in a trading station near Baraka. When he awoke the next morning, someone named Walker was waiting to meet him. This wasn't the R. B. N. Walker whom he'd read about in London but the elusive Reverend William Walker himself, the man who'd known Paul for more than a decade and who had provided him with a home base during his expeditions.

Walker took one look at Reade—sleeping next to his piles of supplies, covetously safeguarding them, fearful of being robbed—and decided he had to offer him a place to stay. It wasn't so much an act of generosity as one of convenience: Walker was certain that the inexperienced Reade would run into problems, most likely conflicts with natives, and as a fellow white man Walker would be the one responsible for bailing him out of trouble. Keeping Reade close at hand would make intervention easier.

Walker didn't tell that to Reade. "As you have not been in a hot country before, you will be sure to have a bad fever," Walker said, "and it will be more convenient for me to attend you in my house than here."

Reade instantly took a liking to Walker, who in turn viewed Reade's naïveté and irrepressible urbanity as sources of mild amusement. "He is very social, & likes our simple way of living, though he has an annual income of seventy thousand dollars," Walker wrote to a friend days later, unimpressed by the information that Reade had revealed. "But income will not keep off fever, nor cook a fowl. So he is just on a level with us all; only when he chooses, he can go home."

Walker was an understated man, and he had a melancholic streak that he was never able to shake after his wife died in the early 1840s. It wasn't his style to get mixed up in a controversy like the one that had enveloped Paul. His silence in the London press was more a reflection of a determination to mind his own business than an indictment of Paul. To Reade, Walker spoke kindly of the young man he'd met back in 1848. But he told him that he couldn't verify all he'd written in his book, because he hadn't accompanied him. That said, Walker told Reade he sincerely believed that Paul had done exactly what his critics were alleging he hadn't: he'd hunted gorillas, observed them in the wild, and shot them himself.

Other missionaries in Corisco and Baraka told Reade the same thing. They also told him that Paul was a crack marksman, a knowledgeable naturalist, and uncommonly courageous and had silently endured more privations and challenges during his travels than he had mentioned in his book.

Reade spent several weeks traveling up the Muni River and then headed south to the Fernan-Vaz region, where most of Paul's encounters with gorillas had allegedly taken place. Exasperatingly, none of the natives seemed immediately willing to take Reade gorilla hunting. Reade had hoped to verify or disprove Paul's claims by repeating his feats.

The days dragged on, and not only had Reade failed to shoot his own gorilla, but he hadn't even *seen* one. On April 20, he wrote to Reverend Walker, suggesting that he felt tremendous pressure to send an initial progress report to London and that he feared his trip was proving a failure.

"I *must* shoot a gorilla, and have something to say about these cannibals," he wrote.

In the Fernan-Vaz region, Reade lucked upon a young interpreter named Mongilomba who informed him that he'd traveled with Paul during part of his journey. When Reade questioned him about the gorilla hunts, Mongilomba recognized Reade's eagerness: R. B. N. Walker had quizzed him about the very same subject months before. Mongilomba told Reade that he'd told Walker that he'd been with Paul when he shot two of the gorillas. But now the young interpreter told Reade he had lied to Walker. He never witnessed those slayings himself, he admitted.

This put Reade on high alert. Days later he met the leader of a local tribe named King Quengueza, whom Paul had praised in *Explorations and Adventures* as intelligent, sensible, and brave. Through an interpreter, Reade asked him if he believed that Paul had really hunted gorillas. The old king didn't hesitate.

"His answer was precise," Reade later wrote. "He and Paulo (as he called M. Du Chaillu) had been in the habit of shooting gorillas together in the bush."

But the next day, Reade met the wife of another chieftain. Unlike Quengueza, she spoke English. Although she hadn't met Paul, she'd heard rumors suggesting that he hadn't shot any gorillas himself.

"He had only shot little birds," she told him.

Reade was intrigued. To him, this woman seemed to have "an intelligent face." She seemed more trustworthy than Quengueza's people, whose black skin he described as "the color of disease" and who constituted a degraded race that "imitates the white man as the ape imitates the negro." When Reade told her that Quengueza claimed to have shot gorillas with Paul, she quickly brushed it aside.

"Ah, you must not believe all these people tell you," she said. "They do not speak the truth."

Quengueza happened to be standing nearby. He and the woman began to converse in their native language.

"What's he saying?" Reade asked his interpreter, unsure of what was going on.

"He says that Paulo and he went a long way into the bush."

Reade decided to try to clarify the confusion by appealing directly to Quengueza, eliminating the filter of the interpreter. Reade held his arms out, trying to pantomime the act of holding a rifle, and he uttered the word the natives used for gorilla: *"njena?"*

The king shook his head from side to side, as if to say no.

Reade interpreted this as a confession. The old man, Reade decided, had never hunted with Paul.

Later, when Reade asked other villagers to confirm that gorillas beat their chests with their hands, as Paul had described in his lectures and his book, the men laughed. That was just a ridiculous story that Quengueza had fed to Paul, they said.

On September 7, Reade sent a letter to the *Athenaeum* declaring that he had finished his journey. The letter unfolded like a juicy exposé of untruths. He revealed that his assistant, Mongilomba, was merely a young boy and "not a gorilla hunter at all" when he'd been with Paul. Reade's verdict was definitive.

"Having spent five active months in the Gorilla Country," Reade wrote, "I am in a position to state that M. Du Chaillu has shot neither leopards, buffaloes, nor gorillas; that the gorilla does not beat his breast like a drum; . . . that the young gorilla in captivity is not savage; and that while M. Du Chaillu affects to have been 'a poor fever-stricken wretch' at Camma (June 1st, 1859), he was really residing in robust health."

Near the end of his letter, Reade conceded that many of the locals held Paul in high regard and that some of the book's descriptions—including those of the Fang cannibals—were "very good," though occasionally exaggerated. But that was faint praise. Reade had already dismissed most of the book as a journey into the land of make-believe.

In refuting Paul's descriptions of gorillas, Reade had zeroed in on the accounts of the animal's supposed tendency to beat its chest when threatened. Reade reported that Quengueza was the source of this misinformation. Paul had never witnessed it himself.

"Thus in an obscure African village an old savage could tell a lie, which has blazed through Europe," Reade concluded.

CHAPTER 34

The Wager

Paul was incensed when he read Reade's report—understandably, because it was unfair.

How could someone who spent only a few months in Gabon, and who had never once even caught a fleeting glimpse of a gorilla, confidently claim more knowledge of the animal than he could?

He grabbed a pen and dashed off an angry letter to the editor.

"What are five months to traverse and explore the immense tracts of country I visited, to ascend rivers, scale mountains, become intimate with numerous tribes, learn to understand their various dialects, occupy many days in palaver with each in succession, make the stay with each which is necessary to acquire their confidence, and pass on to others; and how can any one with such a limited experience only venture to contradict such experiences as mine?"

Reade believed that his own failure to hunt a gorilla proved the difficulty of the enterprise, and therefore cast additional doubt on Paul's claims. But Reade's failure, in Paul's eyes, was easily explained: he hadn't won the hard-earned trust of the natives. Reade never spent more than a few weeks with any one tribe. He didn't speak their languages. In fact, he openly *despised* them as treacherous fools.

"The old African hunters will not take you to the gorilla's haunts unless they have confidence in your nerve and skill," Paul wrote, "and Mr. Reade did not allow them sufficient time to acquire this confidence, if his nerve and skill fully entitled him to it."

The supposed revelation from Mongilomba—the young interpreter whom Reade triumphantly exposed as "not a native hunter at all"—was a perfect example of Reade's distorted sense of truth telling. Paul had temporarily employed Mongilomba, then just fourteen years old, as a camp helper, which he had mentioned in his book. He had never implied that the boy was a hunter or that he had ever accompanied him into the forest to hunt.

Yet Reade's bold declaration that Paul had never shot a gorilla, or any other large animal, was based on exactly this sort of evidence.

At wit's end, Paul issued a challenge to those who believed he'd invented the stories in his book: he bet them that he could repeat the feat. They were welcome to accompany him on a return journey to Gabon, and there they could observe him directly interacting with the tribes, hunting gorillas, and preserving the specimens. All they had to do was to help pay for the trip if he was proved right. He wrote to the *Times*:

> I will prove how I got my specimens, by the simple
> expedient of hunting and shooting others as I hunted and shot
> these, under certain conditions, which involve a fair offer. If
> Dr. Gray and his friends will raise among them and deposit
> in a bank 2,000 pounds, I will deposit 1,000 pounds on
> my side. I will start for the gorilla country, and if I do not

*kill five or six gorillas in the course of two years (I allow
so long for fevers and other contingencies), and bring their
skins and skeletons home preserved with a preparation which
these gentlemen shall give me (provided it is ascertained to
be suitable for the purpose), I will forfeit to them my 1,000
pounds. On the other hand I shall claim their 2,000 pounds
to repay my expenses if I succeed, and I shall be very happy of
the company of the bravest of them in making the venture.*

None of them took him up on the wager, but their refusals did nothing to restore Paul's credibility with the public. By declining his offer, they stripped Paul of his only means of definitively proving his case.

Desperate, he sold his only possessions of real value: the gorillas and other stuffed specimens that had made him famous. Owen, who continued to support Paul in the face of the criticism, arranged for the British Museum to pay £500 for some of them. The remainder of Paul's collection was taken to an auction house in Covent Garden. The skins and skeletons were put up for public bids, one at a time. The most sought-after gorilla skin, an adult male, sold for £110—"a sum much below that at which it was valued by the owner," according to the *Times*. The other skins and skeletons sold for between £4 and £20 each. "The prices realized by the sale of the birds were too insignificant to notice," the newspaper reported.

It had been eight years since Paul had first risked his life to pursue a beast that had never been faced—a risky gambit to a future where he could be judged on his own merits, not tethered to his ancestry. He had guarded secrets, spread white lies, struggled to attract the broadest audience possible—all to protect the possibility of that future. But now the public judged him a fraud. His determination was dismissed as mere tinsel and sham.

Paul's celebrity, quickly souring to infamy, had spun out of his control. It was true that some of his descriptions had been colored by imagination. Mistakes had arisen from his lack of expertise, and he'd grasped at fame too eagerly. Like Reade, Paul had thought he'd found a shortcut to glory—but he swore to his friends that the path itself hadn't been a fanciful dream. He truly had traveled to the places he said he'd gone, and he swore by his descriptions of the gorillas in the wild. But if he couldn't prove it, every gamble he'd ever risked would be lost.

Near the middle of 1863, he counted the money he had earned from the sale of his specimens. It wasn't as much as he'd hoped—the equivalent today of about eighty thousand dollars.

He decided to wager it all on the only thing he believed he could regain control of: himself.

Part Three

CHAPTER 35

The Reinvention

Paul decided to return to Africa. Instead of proving himself by killing more gorillas and bringing them back, he hoped to vindicate himself by carefully compiling the kind of irrefutable evidence and data that he lacked the first time around and that had left his entire enterprise vulnerable to doubt. This time, he'd chart his routes on precisely plotted maps, keep records of astronomical data, and chronicle every step. Instead of being content in the role of an adventurer, he wanted to transform himself into a bona fide explorer, a title that his critics insisted he didn't deserve. During much of 1863, he prepared for the expedition with a disciplined program of self-improvement.

How could someone without any scientific or geographical training become a respectable Victorian explorer? In the mid-nineteenth century, an entire literary subgenre helpfully sprang up to answer this question. The Royal Geographical Society not only housed many of these how-to-explore books in its library but also sponsored the publication of several of them.

In the 1850s the RGS produced a field manual called *Hints to Travellers*, which compiled tips from many of its most far-flung voyagers. The handbook was "addressed to a person who, for the first time in his life, proposes to explore a wild country, and who asks, 'What astronomical and mapping instruments, and other scientific outfit, ought I to take with me? and what are the observations for latitude and longitude on which I should chiefly rely?' To this end we give a list of instruments, books, and stationery, complete in itself, down to the minutest detail, so that an intending traveller may order his outfit at once."

Francis Galton, the organization's secretary, a year later expanded this theme with *The Art of Travel; or, Shifts and Contrivances Available in Wild Countries*. A glance at the index hints at the breadth of the book's ambition (see entries like "bones as fuel" and "savages, management of"). Other exploration guides included *What to Observe; or, The Traveller's Remembrancer*, written by Colonel Julian R. Jackson, another former RGS secretary. A new edition of Jackson's book had just been issued in 1861.

"With such a book in his hand," wrote Randolph Barnes Marcy, an American army officer whose *Prairie Traveler* was must reading for rough travel on both sides of the Atlantic, "[the reader] will feel himself a master spirit in the wilderness he traverses, and not the victim of every new combination of circumstances which nature affords or fate allots, as if to try his skill and prowess."

All the books emphasized the importance of detailed, systematic observation. Along with advice on fording rivers and organizing campsites, readers were given detailed how-tos on geological surveying and orienteering, equipment checklists, and advice on what exactly to look for when judging the physical and moral health of native populations. Richard Burton, for one, rarely traveled without such books; he even edited the

British edition of *The Prairie Traveler*, footnoting Marcy's text with his own observations. And it was Burton's influence that helped prevent the RGS from distancing itself from Paul at a time when some of its members seemed to have given up on him; even Murchison, one of his most energetic backers, had been uncharacteristically silent as the most recent controversy developed.

Shortly after Burton moved to West Africa to assume his consular duties, he had organized a trip to Gabon to search for gorillas—not a traditional activity for a consul, to be sure, but Burton was anything but a traditionalist. At first, his trip, undertaken around the same time as Reade's, seemed to cast more suspicion on Paul's stories. R. B. N. Walker reported to the *Times* that Burton had written to him personally to say he had detected "gross errors" in Paul's account. But after he saw Walker's statements in the newspaper, Burton responded with his own letter to the editor saying that Walker had unfairly represented his conclusion.

"I will briefly state that, after a residence of about three weeks in the Gaboon country, during which I walked to Cape Lopez and explored the south-eastern fork of the river beyond any former traveler, my opinion of M. Du Chaillu's book is higher than it was before visiting the land of the gorilla," Burton wrote. "The Mpongwe natives give 'Mpolo'—i.e., the 'big man,' their corruption of M. Paul's name—the highest character as a *chasseur*. No one, save the jealous European, doubts his having shot the great anthropoid (mind, I modestly disbelieve in the danger)."

Burton's vote of confidence didn't change the minds of many of Paul's critics, but it preserved just enough credibility among the learned men of London to give the would-be explorer their blessing in his quest to redeem himself. With their encouragement, Paul hit the how-to books. But he did more than simply *read* them; with the support of the RGS, he sought out the authors and editors of those books for direct, hands-on instruction.

SIR GEORGE BACK, a sixty-six-year-old British navy vice admiral, was busy revising *Hints to Travellers* for a new 1864 edition. In his younger days, Back had served under Sir John Franklin during his pioneering expeditions through Canada's Northwest Territories. Later, Back had led his own exploratory surveys of the Arctic. As a result, he was an accomplished navigator and handy with all the state-of-the-art instrumentation needed to determine his precise position on the globe. Paul developed a quick and lasting friendship with Back, who tutored him in the basics of geographical science.

In *Explorations and Adventures*, Paul had claimed to summit a peak he called Mount Andele at a time when Reade said he was relaxing on the Atlantic coast. He had written that the mountain stood in the heart of a region that many of his critics claimed he'd never visited. "We were two days about the ascent," Paul had written, "which was a tedious affair, and without its reward, as, when I reached the summit, I found it enveloped in clouds, and mists, and forests, and could get no view at all." The lack of concrete detail fed suspicions that he'd invented the incident. Why hadn't he provided altitude measurements of the peak and those of the surrounding range so that geographers could verify his route? Because at the time he'd had neither the know-how nor the instruments to take such readings.

Paul vowed that this time he'd do things right. Back taught him how to use instruments, such as aneroid barometers, to determine his exact altitude at any given time. Together, they visited the finest watchmakers and meteorological technicians in London, and Back recommended custom-made instruments for Paul to take with him. To protect their sensitive calibrations, he even provided Paul with the same kinds of water-resistant leather carrying cases that he'd used in the Arctic.

As he studied with Back, Paul also took an advanced course on astronomy, which was a requirement to precisely gauge geographical coordinates. He convinced the head of the Royal Observatory in Twickenham to personally tutor him. To ensure that he could use those coordinates to draw accurate maps immune to criticism, he studied cartographical technique with the RGS's map curator, the man who had authored the mapping and orienteering sections of *Hints to Travellers*.

To erase doubts about his descriptions of gorillas in the wild, Paul had another plan: he'd take pictures of them. Thomas Malone, the photography expert who had exhausted his patience during the Ethnological Society meeting, had held Paul to an unreasonably high standard in demanding photographic evidence of him. In the mid-1850s, practically no explorers took pictures in the field. The cameras, chemicals, and glass plates were far too cumbersome and delicate to survive extended overland journeys, and few explorers had the expertise necessary to use them. But photographic technology had advanced considerably in the years since Paul's first encounters with gorillas, and he resolved to become an expert in expeditionary photography.

Paul didn't study with Malone, who had become one of the most prestigious photography instructors in London, but he found another teacher whose credentials couldn't be surpassed: Antoine Claudet. The French-born photographer had learned the art of the daguerreotype from Daguerre himself in the 1830s, and Claudet even owned part of the patent to that process. The photographer's subsequent inventions (the red-filtered darkroom light, among others) had helped refine the emerging art form. He had even been appointed as Queen Victoria's royal photographer. In his studio on Regent Street, the sixty-five-year-old master taught Paul everything he'd need to know to capture images in the jungle—provided his photographic subjects stayed still. Paul learned how to mix chemicals, how to manage tricky outdoor exposures, and how to properly handle the glass plates (film had not yet been invented) during development.

As he collected more skills, he discovered that he needed more and more equipment, all of which would have to be hauled overseas. It was a logistical challenge he hadn't faced during his first expeditions. Large trading vessels regularly traveled between Liverpool and the mouth of the Gabon River, but after that initial voyage Paul would still need to move all of his stuff to the Fernan-Vaz Lagoon, some two hundred miles south, to enter the forest. Ships of sufficient size rarely followed that route. So he was forced to charter his own.

ON AUGUST 6, 1863, a hundred-ton schooner called the *Mentor* was tied to St. Katharine's Dock in London. For several hours, Paul paced the pier, watching as one tin chest after another was hauled onto the ship. They contained almost everything he'd purchased from the sale of his specimens.

Some of the boxes were packed tight with dozens of pairs of balmoral lace-up hiking boots, linen camp slippers, light-colored flannel shirts, pocketed jackets, thick cotton pants, and leggings. He had hundreds of pounds of gunpowder, dozens of rifles and revolvers, thousands of bullets, gallons of castor oil, quarts of laudanum, and enough quinine to treat a small army. He was also taking a weighty store of arsenic, watches, watch keys, sextants, binoculars, a telescope, a sundial, brass aneroid barometers, prismatic and pocket compasses, drawing pens, protractors, thermometers, lanterns, candles, magnifying glasses, rain gauges, almanacs, journal books, skeleton maps, matches, and flints. He had seven pounds of mercury in a stone bottle to create artificial horizons and measure the reflective angles of stars. He had nautical almanacs for the next four years. He had a magnetic-electro machine with a ninety-foot cord for conducting wire, glass tubes and jars to collect insects and worms. He had enough photographic chemicals and equipment to make two thousand pictures, and this alone filled ten boxes. In all, he had fifty-seven large chests of equipment, plus fifty "voluminous bundles of miscellaneous articles," including everything from thousands of pounds of beads for trading to several Geneva music boxes that might be used to impress the natives.

Before Paul left London, John Murray had given him fifty pounds to buy presents for the local chiefs of the Fernan-Vaz region. "Be assured that I will apply the amount in a way that will be always an agreeable souvenir to myself and which I think will please you," Paul wrote to Murray after he received the money. The resulting shopping spree bought enough stovepipe hats, coats, umbrellas, and silk finery to fill a large trunk.

The voyage south along the west coasts of Europe and Africa was long but uneventful. On October 8, 1863, after stops in Accra and Lagos, Paul spotted the verdant coast near the Fernan-Vaz Lagoon.

There, beyond the palms and mangroves, he hoped to

wander inland armed with his shiny new gear. He didn't know exactly how far he would travel or how long the expedition would last. The objective, in his words, was "to fix with scientific accuracy the geographical positions of the places I had already discovered, and to vindicate by fresh observations, and the acquisition of further specimens, the truth of the remarks I had published on the ethnology and natural history of the country."

More than gorillas, he was hunting for validation, and he didn't want to return to London until he found it.

⟞⟐⟝

CHAPTER 36

Damaged Goods

The *Mentor* lingered a teasing distance off the Gabonese coast for days, unable to move closer to the shore. The surf was too rough. Impatient, Paul lowered himself into a canoe and paddled through the waves to survey the mouth of the Fernan-Vaz Lagoon, trying to find a suitably tranquil place to disembark. But an incessant progression of seething whitecaps, in combination with a tricky maze of shifting sandbars, proved too risky for the *Mentor*. The ship would have to remain at sea until the surf calmed, if that ever happened. Unwilling to wait, Paul enlisted the help of natives. He decided to unload all of his equipment into their canoes and trust the Gabonese oarsmen to paddle straight through the breakers, all the way to the sandy beach.

It took the crew hours to transfer his cargo into the bobbing pirogues, which sank a little further into the water as each new crate was carefully lowered inside. One canoe was filled with guns, another with ammunition. The gifts he'd bought for the local leaders were dropped into another, along with the boxes of clothes and shoes.

He saved his most precious cargo for the very last canoe, which bumped up against the *Mentor*. In went the sextants, chronometers, prismatic compasses, and medicine. To make sure that these items were safe, Paul and Captain Vardon, who helmed the *Mentor*, would ride along in this craft with a few native oarsmen.

The other canoes departed first, pushing away from the ship and nosing toward the beach. Paul watched the men paddle away from him, the canoes seesawing over the waves, dipping in and out of sight, jouncing and splashing.

Under his light coat, Paul wore a life vest made of cork. He carefully stepped down into the last canoe with Vardon. They settled down among the boxes and crates as the oarsmen guided the wooden vessel into the waves, which seemed to be gathering force behind them.

Their strategy was to reach land as a surfer would: race to get on top of a surging wave, and let it carry them home. Paul peered behind him, and he saw a powerful swell approaching. The oarsmen took off, paddling hard. The wave was gaining on them. They stole quick glances over their shoulders to monitor its progress, then paddled even harder—but they couldn't paddle fast enough. The wave crashed over the canoe, hurling Paul into the sudden silence of the water. Struggling to collect his bearings, he popped back to the roiling surface. He gasped for air between a relentless succession of breakers that kept pushing him back under. The cork vest couldn't fully counteract the downward tug of his soaked clothing. As both he and Vardon fought to keep their heads above the waves, some of the oarsmen reached them to help. They wrestled the coat off Paul and "with great exertions kept me from sinking."

A group of onlookers standing on the shore had seen them struggling and dispatched rescue canoes. But the powerful swells held the paddlers back. The men in the water battled exhaustion. When the waves finally subsided, a rescue canoe arrived and someone yanked them out of the water. Paul and the others were breathless with fatigue—but safe.

The canoe's precious cargo didn't fare so well.

Some of the boxes, tossing in the waves, were collected by the natives. Others washed ashore later. Some of the pieces of equipment remained undamaged, but most were ruined. For the moment, so was the primary objective of Paul's expedition.

Dejected, Paul dried out the waterlogged instruments, their delicate gauges splayed to meaningless readings. He wrapped them up and shipped them up the coast on a small boat to Baraka, where they were in turn shipped back to the stores in England where he'd bought them. By mail, he pleaded with the shop owners to repair, or even replace, the damaged goods.

Most of the shop owners agreed to help him, but all Paul could do now was wait for the replacements, which would arrive a full nine months later. But it was not time wasted.

EACH DAY that he waited, his expenses mounted. To help defray the costs, Paul came up with a plan to help Captain Vardon fill the *Mentor*'s hold with goods to sell back in England. The process took about four months, but it was worthwhile. Paul used his experience dealing with native traders to help him acquire ebony, palm oil, and barwood. In return, Vardon reduced the debt that Paul owed him for chartering the ship. Paul also earned a lot of goodwill from the tribal leaders, who were thankful for the business.

The unplanned pause also forced Paul into a relaxed pace that afforded him time to play with some of the cameras and astronomical instruments that had survived. With a clear view of the horizon and well-mapped coordinates, the coast was a perfect place to experiment with his sextants. "I like to practice while I am near to the sea-shore," he wrote to Murray during his delay, "as I know exactly my longitude, and would then know if I am correct."

His unhurried attitude seemed to put the Nkomi, the tribe that inhabited this stretch of the coast, at ease. He paid calls on friends he'd met years before, and he earned the trust of the same men he'd have to depend upon as porters when he finally headed inland.

King Quengueza, the chieftain who had allegedly told Reade that Paul had lied about his gorilla hunting, lived in a village about eighty miles upriver, but Quengueza traveled to the coast to greet his old friend as soon as he heard Paul had arrived. Reade had portrayed the king as a pathetic caricature of ersatz royalty, but Paul didn't share that opinion. "I felt and still feel the warmest friendship towards this stern, hard-featured old man," he wrote, "and, in recalling his many good qualities, cannot bring myself to think of him as an untutored savage."

The reunion was warm but cloaked in the strict protocol of a formal royal greeting. Gifts were ceremoniously exchanged. In front of a group of royal hangers-on, Quengueza presented Paul with a goat. In return, Paul presented the king with a copy of his book (the illustrations, he knew, would be cherished) and a set of silver dining ware, which he had purchased in London with some of the money Murray had given him. "The old fellow was so delighted," Paul wrote to Murray soon after the exchange, "that he said that he would send you a very large stick of ebony."

Paul's most striking gift to Quengueza was a coat that

he'd had specially tailored for him at one of the finest clothiers in London. It was bright blue, with garish yellow trim and red lining. The king was duly impressed, as Paul knew he would be, but there was more. Paul dazzled him with a chestful of silks, cottons, gunpowder, flintlock rifles, and enough beads to bejewel every one of his wives.

It was a payoff, plain and simple, and Paul expected protection in return. Quengueza responded exactly as planned. He promised that when Paul ventured inland, he could count on the services of as many men as he needed. The king personally guaranteed their loyalty.

WHILE HE lingered on the coast, a group of native hunters found Paul in his encampment and told him they'd captured something he might be interested in buying: two live chimpanzees, a male and a female. Chimps weren't nearly as prized as gorillas, but they were still rare, particularly living specimens. Paul bought them, trained them, and kept them around his camp as pets. He named them Tom and Mrs. Tom.

When Vardon was finally ready to take the *Mentor* back to England early in 1864, Paul coaxed the two chimpanzees into boxes and loaded them onto the ship, along with a three-month supply of bananas. Mrs. Tom died during the ocean voyage, but the male survived.

The chimp was taken to the Crystal Palace in Sydenham and became one of the most popular zoological attractions in England. He lived there for two more years but died, horribly, in a fire that destroyed the palace's north wing in 1866. According to newspaper reports, workers fleeing the building reported hearing Tom's "frantic cries" as the frightened chimpanzee clutched the hot iron bars of his cage, unable to escape.

Paul couldn't have predicted the animal's tragic end. As far as he knew at the time, his transfer of a living ape from Gabon to London was an unqualified success. He hoped to repeat the feat with a gorilla.

In addition to the live specimens, Paul was collecting an even odder sort of contraband to be shipped to London: human skulls. He wanted to send them to Owen and to his other friends at the Anthropological Society, to be measured and compared with the skulls of other racial groups. Owen believed such skulls might help demonstrate the differences—not similarities—between men and apes.

Paul told the Nkomi tribesmen that "there was a strong party among the doctors or magic-men in my country who believed that negroes were apes almost the same as the gorilla, and that I wished to send them a number of skulls to show how much they were mistaken." As incentive, he said he'd pay three dollars for each skull they found.

Soon, he was overwhelmed with skulls scavenged from native burial grounds. When he had collected more than ninety, he was forced to reduce his price.

PAUL ROSE at dawn to travel to a village on the eastern bank of the Fernan-Vaz after reports that a band of gorillas had been seen on a small plantation nearby. When he and a young man named Odanga arrived at a clearing of manioc plants where the gorillas had been spotted, all was quiet. But as Paul walked alongside the plantain trees that fringed the clearing, he heard a leafy rustle of movement.

Paul hid behind a bush, keeping still. The rustling resumed, and soon he caught a glimpse of a female gorilla. He shifted to get a better view, and two more emerged. Then a fourth. None of the gorillas saw him. He remained frozen in place, awed but completely safe, careful not to destroy the best opportunity he had ever had to observe the apes in the wild. If he wanted to confirm the descriptions he had included in his first book, this was the way to do it. He watched in silence, without raising his gun and blowing his cover.

They tore into the shrubby plantain trees, yanking the stalks with their powerful arms. Ripping apart the base of the tree exposed the juicy interior of the stalks, which they devoured greedily. Some of the animals made an odd sort of clucking noise as they ate, which he assumed signaled contentment. Every so often, the gorillas looked up from their meals and scanned the landscape, but they didn't appear to see him behind the bush.

"Once or twice they seemed on the point of starting off in alarm," he wrote, "but recovered themselves and continued their work."

Gradually, as the gorillas worked their way through one plant after another, they slipped out of view and left the plantation.

Paul spent the night in the village. The next morning, as he was crossing a deep hollow planted with sugarcane, he was surprised to see an enormous gorilla on the opposite slope, staring straight at him. Paul was without a shotgun, armed only with a small pistol, which likely wouldn't have killed the animal. He was scared; it was the first time he'd faced a gorilla without a rifle. Normally, he would have quickly jumped to raise his gun. Now he was forced to adopt a passive attitude of stilled shock.

"The huge beast stared at me for about two minutes," he later wrote, "and then, without uttering any cry, moved off to the shade of the forest, running nimbly on his hands and feet."

Without gun in hand, he was able to see the gorillas with clearer eyes than ever before. He took careful note of the way they walked, noticing how the arms were almost completely straight when they touched the ground—not bent or bowed as they had been pictured in some of the illustrations in his own book.

·◦}[ ]◦·

CHAPTER 37

The Boldest Venture

After sending the live chimpanzees to London, Paul spread the news among all the neighboring tribes that he would pay handsomely for a live gorilla. To his shock, a group of tribesmen quickly rose to the challenge and executed the most successful gorilla roundup in history. They brought him three live gorillas: an adult female, a tiny infant, and a very young male, not yet fully grown.

In *Explorations and Adventures*, Paul had written that capturing an adult gorilla would be impossible. Now he realized he should have added the qualifier "unless wounded."

The capture had been effective but brutal. The hunters had stumbled upon a group of female gorillas and their young near the Fernan-Vaz Lagoon. The silverback male was not among them, so the men were braver than usual. Armed with guns, axes, and spears, the hunters formed a line and forced the gorillas toward the water. Sensing the danger, the gorillas panicked. Several of them escaped in the ensuing melee, but the hunters managed to seriously wound one adult female with a bullet to the chest, and they beat down a juvenile male with clubs. The infant was defenseless.

Paul examined his new acquisitions with a mixture of

awe and pity. Whenever he stepped toward the young male, the animal would rush toward him and then pull up short of Paul and beat a hasty retreat. "If I looked at him he would make a feint of darting at me," he wrote, "and in giving him water I had to push the bowl towards him with a stick, for fear of his biting me."

The adult female, however, was in no shape to intimidate anyone. Her chest wound was critical, and her head had been pummeled by the clubs when the hunters bound her. One of her arms was also badly broken.

After she spent one agonizing day in captivity, her groans diminished to faint whimpers, then tapered into silence. The infant hugged her breast, trying to coax milk from her. Taking advantage of their stillness, Paul set up his camera and began the painstakingly difficult process of trying to capture a discernible image on glass plates. Eventually, he took a portrait of the two gorillas. But the mother was dead, and the baby clung to her in sad desperation.

"Her death was like that of a human being," Paul wrote of the mother, "and afflicted me more than I could have thought possible."

He tried to nurse the infant on goat's milk, keeping it alive for four days.

"It had, I think, begun to know me a little," he wrote.

The surviving young male was still in good health when Paul's new and repaired instruments finally arrived in mid-1864. Just as he'd done with the chimpanzee, he penned the gorilla in a box and gave the captain of a ship an ample store of bananas to feed the animal. As incentive, Paul promised the captain an extra hundred pounds if the gorilla made it to London alive. Paul even guaranteed the cargo by registering it in care of a maritime insurance firm. "I had sent him consigned to Messrs. Baring, who, I am sure, never had any such consignment before," he wrote.

Paul watched the canoe carrying the gorilla get drenched by breakers, and he could see that the dousing only made the animal more angry. Paul hoped that fiery spirit wouldn't be entirely dampened by the long journey on the steamer; he wanted his friends to see the animal with its aggression intact. If a gorilla jumped out of that box dockside in England, blind with panicked rage, Paul knew none of those spectators would again accuse him of exaggerating the gorilla's menace.

BEFORE HIS instruments arrived, Paul wrote a letter to John Murray requesting a copy of the book *Cosmos*, by Alexander von Humboldt.

Humboldt had been an invisible guiding hand behind all the how-to-explore books that Paul had studied. The German explorer had journeyed widely throughout Central and South America between 1799 and 1804, and his *Personal Narrative of Travels to the Equinoctial Regions of America* had set the standard for exploratory narratives. As a meticulous observer of natural phenomena, Humboldt was the model for generations of scientific travelers, and handbooks like *What to Observe* and *The Art of Travel* repackaged many of his ideas to advise would-be explorers exactly where their attentions should be directed.

Cosmos was Humboldt's ultimate statement, a manifesto celebrating the essential interconnectedness of the universe's natural processes. He published the first volume of the five-volume work in Germany in 1844, and it wasn't released as a complete edition in English until 1858. Drawing on Humboldt's lifetime immersion in scientific fields that ranged from geology to botany, the book's driving idea is that the natural world is wholly relational: a "chain of connection, by which all natural forces are linked together, and made mutually dependent upon each other." The highest calling of man, he believed, was to perceive the existence of these connections.

Humboldt was full of ideas that would have made a strong impression on Paul, though some of them directly challenged the worldview that he'd subscribed to during his first expedition. Humboldt didn't particularly like sportsmen, and in his book on South America he disparaged the "troops of marauders, who roam over the steppes killing the animals merely to take their hides." With the publication of *Cosmos*, Humboldt was advancing ideas that presaged the ecological movement of the next century. Around the same time that Paul read the book, it was helping to inspire a burgeoning realization that mankind could permanently and adversely alter nature by altering the chain of connections. Ralph Waldo Emerson and Henry David Thoreau, who had read parts of *Cosmos* as early as 1845 and 1850, respectively, helped spread Humboldt's influence among English-language readers, but none reinterpreted his ideas as forcefully as George Perkins Marsh, whose 1864 book, *Man and Nature*, would prove to be the seminal document for the conservation movement. Marsh suggested that when human beings targeted certain species for slaughter or destroyed a part of the environment, it could result in consequences no one could have foreseen.

These were novel ideas for the time. Like almost everyone else, Paul had always treated nature as inanimate clay meant to be molded by man. But as he prepared for this new expedition, different ideas were infusing the world of exploratory science. He was by no means in the vanguard of promoting them, but he couldn't help but notice that the prevailing wisdom concerning the killing of animals was changing. His personal objectives reflected it.

"It was not my object on the present journey to slaughter unnecessarily these animals," he wrote, "as the principal museums in civilized countries were already well supplied with skins and skeletons, but I devoted myself, when in the district inhabited by the gorilla, to the further study of its habits, and the effort to obtain the animal alive and send it to England; hoping that the observation of its actions in life would enable persons in England to judge of the accuracy of the description I gave of its disposition and habits; at least to some extent, as the actions of most animals differ much in confinement from what they are in the wild state."

The statement of purpose was open to interpretation. He undoubtedly desired to elevate his enterprise above that of a mere sportsman, but it also served to disarm anyone who might accuse him of failure if he didn't bring back the same kinds of specimens he'd collected years earlier. Regardless, this much would be certain: for the rest of his life, Paul would travel the world widely, and he would never again travel as a hunter of wildlife.

"I am perfectly tired of this Gorilla business," he wrote to a friend while he waited on the coast to begin the expedition, "and I intend to have nothing to do with the beast in the future."

WEEKS LATER, letters arrived in Gabon reporting that the young gorilla Paul had shipped to England had died at sea. But Paul wasn't around to receive the news. He'd already departed the coast. His journey, finally, had begun.

Just before he left, on August 20, 1864, he sat down in

his hut and pulled out a notebook, penning a series of letters on those thinly ruled sheets. His mood was somber and reflective. One after another, he addressed notes to many of his closest friends, including the Reverend William Walker, John Murray, Henry Bence Jones, Sir George Back, and Commander C. George at the Royal Geographical Society. To Owen, he sent a shipment of mats, native cloth, a drum, and a harp. He thanked each of them for their kindnesses over the years, and he assured them that he would work hard to reward the faith they'd shown in him. He said he was optimistic, as always, but he clearly recognized that his expedition could kill him. During his first journeys into the forest, he'd been naively unaware of many of the dangers that faced him, and back then there was little pressure on him; he could have simply abandoned his journey if ever it got too rough. This time, he vowed that he would push himself to the absolute limit. To Back, he wrote:

> *In a few days I am off for the interior and I shall push on until I shall find insurmountable barriers that will stop me, and even then I will try with patience and perseverance to go further. I will work hard, and if Him who guideth the steps of man allows me success I shall perhaps reach the Nile. I am in very good health and good spirits and think I will be able to reach very far into the interior, but my hopes after all may be blasted for I am aware of the many accidents which may stop me on my onward cause, disease may get hold of me and may lay me prostrate in a desolate country. I know that I may perhaps die forsaken by all or that I may be killed by treachery. I have thought of all of these things and I have come to the conclusion that I may also succeed. . . . My whole*

*soul is in the work I am to undertake. I pray to have strength
to accomplish it.*

When Roderick Murchison received Paul's letter, he was
struck by the optimism that leaped out from the page. Murchison immediately wrote to Owen: "It is so good, so superior
to his earlier letters, so full of fire, noble and self-sacrificing
resolution, that I shall read it as our opening *morceau* at the
Geographical, November 12. . . . Never were we more in the
right than when we stood up for this fine little fellow."

During the RGS's annual anniversary meeting the following May, Murchison announced for the first time that the
young explorer hoped to travel all the way to the Nile. Murchison told the members that during his delay on the coast,
Paul had already collected and sent back to London numerous
specimens, including thousands of insects that he had painstakingly captured in glass test tubes. The young man that so
many had dismissed as a cavalier fabulist was proving himself
a true man of science, Murchison said.

"In his last letter to me, written on the point of departure
from the coast, he begged me not to be uneasy about him
for a year or two," Murchison said. "As M. Du Chaillu has
rendered himself a photographer, as well as an astronomical
observer—advantages he did not possess in his first journey—
we are sure, if his life be spared, to reap a rich harvest on
his return. And so, let us wish him God-speed by the way!
In boldness of conception nothing in the annals of African
research has surpassed his present project."

CHAPTER 38

The Armies of the Plague

About fifty men accompanied Paul during the first leg of his journey inland. Most expected to be replaced by porters from other tribes deeper inland, but ten were Nkomi tribesmen—the same group that helped during his previous expedition—who promised to stick with him for the duration of his journey, come what may.

Paul outfitted his crew, dressing all the men alike in blue woolen shirts, canvas pants, and red worsted caps. They looked like a traveling army, and in some respects they were. They were issued rifles and assigned specific stations under Paul's command. Most were simple carriers, hauling the outfit in large, plaited cane baskets strapped to their backs. The ten permanent members enjoyed a slightly higher rank, which helped guarantee their loyalty. At some point early in the journey, Paul realized that they believed he was taking them to London, the "white man's country," which lay beyond the jungle and where any man could accumulate unimaginable wealth, just as Paul apparently had.

He failed to correct this misconception.

They traveled first by canoes, then on foot. His most trusted man walked in front, and Paul brought up the rear, ever vigilant in case any of the porters tried to break away from the group with some of the valuable supplies. They trudged through swamps, soaking rain, and clouds of mosquitoes. His stash of Epsom salts dwindled as the feet of his men grew sorer.

He explored the Rembo Ngouyai, a branch of a river he'd discovered on his first journey, borrowing canoes from local tribes, following unmapped tributaries inland, riding out rapids, portaging picturesque waterfalls.

He rarely bothered to lug his shotgun, opting instead for a walking stick and a notebook. Verifiable data was his target, not dangerous beasts. But he quickly discovered that scientific observations could be as elusive as any gorilla.

The overcast skies vexed him. He'd painstakingly pour out a silvery pool of mercury into a tray, creating an artificial horizon; he'd carefully protect the mercury from any breeze, covering it with a glass pyramid. When he was finally ready to capture the reflections of stars and measure their angles, the heavens would cloud over and refuse to clear. When he reached the fabled falls of Fougamou, which he'd heard about during his previous journey but was never able to reach, he was determined to take a photograph to prove that he personally saw them. To get a clear shot, he ordered his men to cut down a tree near the water's edge. By the time the tree fell, a fiendish bank of clouds conspired to sabotage his exposures. At twilight he finally surrendered, putting the camera back in its case without capturing a clear image.

During the first six months of his journey, only two days were free from clouds. Weeks succumbed to the vain quest for empirical proof. By the end of 1864, some of his men had grown impatient with their leader, who seemed more intent on watching the needle of his barometer than moving forward. Even his most trusted Nkomi assistant, a young man named Macondai, struggled to maintain his customary optimism.

"Macondai cursed the *okenda i nialai* (the good-for-nothing journey)," Paul noted, "which did not take us a step nearer to London."

HE ARRIVED in a tribal village called Olenda, which he'd visited years before, and sent word ahead that his party planned to travel eastward through the lands of the Apingi tribe. To Paul's surprise, a message came back almost immediately: he was not welcome.

This harsh denial had roots in his previous journey. The former king of the Apingi tribe and his son both had died shortly after Paul had visited them in the mid-1850s. The people blamed "the white man." He was cursed, they said. The new king didn't want to gamble with his life.

Paul and his men were forced to camp in Olenda, where he worked to reconfigure his route. Days later, before he was done, a young man in the village fell seriously ill. Rumors swirled that the young man, who had carried some of Paul's equipment into the village, had been hexed by the foreign visitor. Within a day, the young man was dead. Around the same time, two more men who had been in contact with the traveling party showed the same symptoms as the dead man: their skin broke out in small red spots, which turned into tiny blisters after a day or two. Then some of Paul's porters fell ill.

It was smallpox, and Paul knew it was deadly serious. He wasn't at risk for infection himself, because he had been immunized in London two weeks before his departure. But the contagion could wipe out a village with terrifying rapidity.

The disease, which can be passed through the moisture of a person's breath, generally takes about twelve days to incubate. Before arriving in Olenda, Paul hadn't noticed any signs of the illness among his men. It was possible that some of the inhabitants of this remote village had already been infected. But it was also possible that members of his group had picked it up after leaving the coast, spreading it into the interior, where the tribes were particularly vulnerable to new diseases.

The villagers of Olenda quickly blamed Paul, who fanned their suspicions when, upon seeing the outbreak, he ordered his men to stay away from areas where the disease had been reported, telling them that smallpox was highly contagious. If he hadn't brought the disease with him, the villagers thought, how did he know so much about it?

"They began boldly to accuse me of having introduced the *eviva* (thing that spreads, i.e., the plague), or, as they sometimes called it, the *opunga* (a bad wind), amongst them," Paul later wrote. "They declared that I had brought death with me instead of bringing good to the people; that I was an evil spirit; that I had killed Remandji, king of the Apingi, and so forth. Hence arose angry disputes."

Within days, more than half of the people in the village had fallen ill, their faces encrusted with hideous pustules, their minds fogged with fever. Paul was forced to search for new porters as his own men began to fall. On three occasions he rounded up enough porters to move his gear out of the village, but each time a crucial number of his crew fell ill before the loads could be fully packed. Seeing Paul's mounting desperation, the healthy men of the village increased the amount of pay they demanded to work.

He was at wit's end. With a skeleton crew of porters, he sent most of his equipment to another village. They were to drop off their loads and return to Olenda to help him and the remaining men carry the rest of the baggage.

While he waited in Olenda, the pox continued to tear through the community. "Not a day passed without its victims," he later wrote, "each fresh death being announced by the firing of guns, a sound which each time pierced through me with a pang of sorrow. From morning to night, in my solitude, I could hear the cries of wailing, and the mournful songs which were raised by the relatives round the corpses of the dead."

Soon there weren't enough healthy men to gather food. The few able-bodied men, including most of Paul's core Nkomi crew, trekked to neighboring villages to petition help, but they were turned away empty-handed. News of the plague had spread as fast as the disease itself. Paul's men were considered a scourge.

The ruler of Olenda lasted longer than most of his subjects. But "King Olenda," as he was called, wasn't immune, and one morning he mentioned that he felt unusually hot and thirsty.

Paul was racked with guilt. He had taken a photograph of the king just days before, and those who were still well enough had gathered around to see the spectacle. Now Paul regretted having pulled out his camera, worried that the "magical" process he'd demonstrated would provide the villagers with yet another reason to be suspicious.

Soon, even Macondai, his most dependable assistant, was prostrate with fever. Paul felt like a pariah, which made the unexpected hospitality that some of the villagers showed him especially poignant. "Those who were now well enough crept towards the plantation to get plantains for me," he wrote, "and even the invalids, men and women, sent me offerings of food, saying, 'We do not want our stranger to be hungry.'"

He had expected hostility, and they had shown him kindness.

Weeks passed slowly while he waited for the men he'd sent ahead to return to help them leave the village. Wherever Paul walked among the locals, he saw hideous images. Men with sores teeming with maggots writhed on the ground. Raving women with graveyard eyes, victims of both disease and near starvation. Flies swarming over the dead. Eventually, Paul noted that the spread of the disease seemed to abate— "from sheer lack of victims for further ravages."

Finally, three of his men from the advance party of porters returned to the camp bearing terrible news: much of Paul's equipment never made it to the next village. Some of the men had simply returned to their remote and scattered plantations, taking their loads with them.

THE RIVERS that flowed down from the mountains were swollen with rain. One of them, the Ovigui, had overflowed its banks, creating three channels separated by slender strips of mud. The only way across was to walk a long and narrow log. Apparently, it was a well-traveled bridge. A rope of twisted lianas stretched across the water just above the log, tied from tree to tree, to serve as a handhold.

When Paul stepped onto the wet wood, his boot slipped. He tumbled down headfirst into a watery hole that was, fortunately for him, disconnected from the main flow of the swift river. As soon as he realized he was safe, he remembered the watches he was carrying. They were wet but unharmed.

This concern for his equipment was perpetual, as was his distrust of the handful of new porters he'd picked up in Olenda. He couldn't afford to lose any more of his diminished supplies, and he'd only known the men for a matter of days. He feared theft. He had no extra clothes. All his sugar and tea were gone. His medicine chest felt light.

His paranoia intensified when some of his instruments—thermometers, aneroids—disappeared as the party trudged deeper into the interior.

THE STEEP mountainsides were crowded with trees that appeared to be lashed together by lianas and creeping ficus. Crystalline waterfalls sprayed a delicate mist over the jungle paths. The men stepped lightly over rotting branches.

"I found it impossible to keep them all together," Paul wrote. "All sorts of excuses were invented for their lagging behind, and I soon made the discovery that they were hiding their provisions in the bush—a sign that they intended to rob me and run away by the same road."

More of his men deserted him. He was forced to send others ahead to another village to try to ask for more help and more food, which was in scant supply. Game animals were hard to find. They survived mostly on kola nuts. His loyal Nkomi men from the coast helped him watch over the others at night in a constant vigil against theft.

By March 24, 1865, he had reached the village of Máyolo. Settled in camp, Paul took an inventory of his gear and found that he'd lost more than he'd thought. His medicine chest had been plundered of castor oil, calomel, laudanum, rhubarb, and jalap. Three thermometers were gone, as were many of his beads. Worst of all, one of his cameras, most of his photographic plates, and his developing chemicals had disappeared. Later he heard that two of his former porters had been found dead back in their villages. It might have been smallpox.

But Paul vaguely wondered if the men might have tasted the highly toxic chemical solutions.

The local villagers had heard of the plague, but the king, who had given the group approval to enter his territory, was more interested in what Paul might be able to offer him in return for the hospitality. Paul laid out whatever gifts he had left—some beads, cloth, a few guns.

"Look!" the king of the village told his subjects, impressed with the gifts. "This is the sort of plague the white man brings among us. Would you ever have had any of these fine things if I had not invited him to come?"

Yet within four days, the king himself was shaking with fever. Paul knew that if the man died, his expedition—and perhaps his life—would be over. None of the tribes in the forest would believe he hadn't brought the plague down upon them. And he couldn't deny it. No matter how the epidemic had begun, Paul couldn't evade responsibility for helping to spread the pox farther inland.

As more people fell sick, a forest fire pressed frighteningly close to the village. As Paul rushed to secure his remaining powder and ammunition, he cursed his bad luck. That night the king continued to moan and wail.

Paul was filled with a nervous, frenzied energy. He scribbled in his journal and made triplicate copies of his notes. He grew obsessed with chronicling exactly what was happening. He instructed his men that if anything happened to him, they should make sure his journals found their way back to the coast, where they could be shipped to England. To the natives, his erratic priorities must have seemed a sure sign of madness.

"On the 1st and 3rd of April I over-exerted myself in taking several solar observations," he reported. "The heat in the shade was about 92° Fahr., and in the sun it reached 130°

or 135° Fahr. I took, at night, several lunar observations, ascertaining the distances between the moon and Venus and between the moon and Spica, and obtained also several meridian altitudes of stars. The sky was so clear that I was anxious not to let the opportunity pass of obtaining these observations. My exertions, however, combined with my heavy anxieties and the loss of my goods, brought on an attack of fever."

He tested the local cure, spiking lime juice with cayenne pepper. The ailing king, meanwhile, required stronger medicine. His pox had reached the ominous blistering stage. A medicine woman rubbed an herbal salve over his skin, and she drew white stripes with chalk down his arms. She chewed roots and seeds, then spit the mushy pulp on his wounds. Finally, she lit a bundle of dry grass and held the flame close, scorching the king's skin from his feet to his head. He only suffered the cure once. His health, somehow, began to improve.

Paul knew that his entire project was in serious danger. After he asked for more porters, the recovering king convened a community meeting to decide his fate.

Like a witness in his own trial, Paul pleaded his case. He reminded the king that he hadn't come for ebony, or to find wives, or to sell goods—only to travel, and he was prepared to reward the local hospitality with gifts. "I told you when I came, and you knew it before, that I wanted to go further away," he said. "Come and show me the road through the Apono country. It is the one I like the best, for it is the shortest. I will make your heart glad if you make my heart glad. I have things to give you all."

King Máyolo slept on it. The following day he announced his decision: he would travel himself to visit an Apono chief who could ensure Paul's safe passage.

But days later, when the king returned from his audience

with the neighboring ruler, he informed Paul that the fear of the plague was too powerful. The tribe didn't want Paul anywhere near them.

On May 14, he scribbled in his journal: "My misfortunes will never terminate!"

Twenty-four hours later, his luck did an about-face: "May 15th. Máyolo's messenger returned to-day with the joyful news that the Apono chief would receive us."

Paul left Máyolo with about twenty men, including his original companions from the coast, except two who were sick with smallpox. Paul carried a forty-pound pack on his shoulders—a considerable burden for someone who, in a weakened state, likely weighed about a hundred pounds.

⋯◃❙❙▹⋯

CHAPTER 39

Running for Their Lives

He had now traveled about four hundred miles inland, farther than any outsider had ever penetrated. When he and his men reached the tiny village of Dilolo—it consisted of little more than a single dirt lane cutting between a couple dozen huts—the entire male population of the village was waiting for him, spears in hand. They had formed a human barricade to keep the infamous spreader of plague out of their town. They had also set fire to a swath of prairie flanking the village, to prevent the oncoming travelers from simply walking around them.

As Paul and his men continued toward them, one of the men blocking the road raised a bow and threatened to shoot. One of Paul's porters responded by leveling a rifle. Instantly, without a word among themselves, nearly all the members of Paul's bedraggled expedition raised their gun barrels in a united front.

Bows and arrows were no match for rifles. The barricade buckled. The villagers could only watch Paul's men pass.

After they marched through, the crew was energized by the victory. Paul heard one of them shout to his compatriots:

"We must go forward. We are going to the white man's country. We are going to London!"

PAUL UNROLLED his tape measure as the old woman flinched. He assured her he wouldn't hurt her. She didn't seem convinced. The tape, stretched from the ground to the top of her head, measured exactly 52.5 inches—or just over four feet four inches.

His group had encountered the woman in a remote swath of forest, and she turned out to belong to a tribe of remarkably small people. Paul, for the first time in his life, felt like a giant. One after another, he approached them with his tape measure, asking them to stand still. The first young man he measured seemed perfectly representative of the men; he stood 54 inches tall, or four feet six inches. No one in the entire village stood over five feet. He measured the circumference of villagers' heads, the distance from their eyes to their ears. He could barely believe what he was seeing, and to buy their cooperation, he gave them beads.

He'd heard rumors during his first expedition of a tribe of unusually hairy "dwarfed wild negroes" who lived somewhere beyond the regions where he'd traveled. But the same people who spoke of this mysterious tribe also insisted there was another cloven-footed race of men that lived even deeper in the forest—a far-fetched bit of folklore that seemed to rob all such tales of credibility. Yet here he was, incredibly, in a squalid wonderland built on a shrunken scale.

They were the first tribe of Pygmies ever discovered in central Africa. Paul wrote down their tribal name as "Obongos," but eventually they became known as the Babongo people. He learned that they were seminomadic. For as long as anyone could remember, they had always stayed within the territory they shared with another tribe called the Ashango but rarely in a fixed place, moving their villages regularly. Paul didn't think they were related to the Ashango: their hair seemed much curlier, and the men had more body hair than the neighboring tribes. His Ashango porters were "anxious to disown kinship with them" but at the same time admired Pygmies' renowned expertise in hunting and trapping wild animals and fish, which they occasionally used to trade with their neighbors.

"My [Ashango] guides were kind enough to inform me that, if I wanted to buy an Obongo, they would be happy to catch one for me," Paul later wrote.

He tried to win the friendship of the Obongo as he did almost every new tribe he encountered: by exploiting the power of technology to appear like a magician commanding mysterious powers. Armed with the music boxes, a large magnet, sulfur matches, and his guns, he put on a show for the natives, many of whom had never seen any article of manufacture more advanced than a glass beer bottle.

> *The musical box was brought out, wound up, and set playing. The people were mute with amazement; at first they did not dare to look at the musical box, afterwards they looked from me to the box and from the box to me, evidently convinced that there was some communication between me and it. Then I went away into the forest, the musical box still continuing to play. When I came back there was still the same mute amazement. The box was still playing, and the people seemed to be spell-bound, not one could utter a word. When I saw that the tunes were played out, I shouted as loud as I*

could "Stop!" and the silence that ensued seemed to surprise
them as much as the music had done before. Then taking my
revolver I fired several times, and my men fired off their guns.

The show helped earn him safe passage through new territories, but his porters weren't so easily fooled. They knew that
he needed them more than they needed him. He continued
to lose equipment to desertion and theft, and it had become
increasingly difficult to keep the men from fighting among
themselves. Days were squandered in arguments. Paul's toler-
ance for the natives, particularly those in his own crew, eroded.
By July 1865, his thoughts had darkened considerably:

> *It is a tremendous task that I have undertaken. The*
> *ordinary difficulties of the way, the toilsome marches, the*
> *night watches, the crossing of rivers, the great heat, are as*
> *nothing compared with the obstacles and annoyances which*
> *these capricious villagers throw in our way. I begin to dread*
> *the sight of an inhabited place. Either panic-stricken people*
> *fly from me, or remain to bore me by their insatiable curiosity,*
> *fickleness, greediness and intolerable din. Nevertheless I*
> *am obliged to do all I can think of to conciliate them, for I*
> *cannot do without them; it being impossible to travel without*
> *guides through this wilderness of forests where the paths are*
> *so intricate; besides, we could not make our appearance in the*
> *villages without someone to take us there and say a good word*
> *for us. . . . I am forced to appear good-tempered when, at the*
> *same time, I am wishing them all at the bottom of the sea.*

His misery wouldn't last long. The expedition that he
originally anticipated lasting as long as five years was about to
confront an insurmountable tragedy.

THEY TRUDGED into a small village called Mouaou Kombo, which lay in the middle of dense forest about 440 miles from the Fernan-Vaz Lagoon, where he'd begun. Paul planned to hire more porters there for the continued march east. But the Mouaou villagers informed him that the occupants of another settlement farther inland had sent word that they would attack this village if Paul were allowed to pass into their territory.

Paul thought diplomacy might work. He sent two members of his party ahead to negotiate with the villagers. But a short time later, four sentries from the other village arrived to inform them that the matter wasn't negotiable. Paul was simply not welcome.

Sensing tension, the Mouaou Kombo chieftain offered Paul refuge in his hut. But some of Paul's porters who remained outside tried to intimidate the sentries by firing their guns in the air. Something went wrong.

According to Paul's account, which is the only one that survived, one of the porters discharged his rifle before raising the barrel to the sky. An errant lead ball whistled through the village. Soon, he heard screams from one of the huts.

"I rushed out, and not far from my hut I saw, lying on the ground, the lifeless body of a negro; his head shattered and the brains oozing from his broken skull," Paul later wrote.

The man who had fired cowered in horror. "Oh, Chaillie, I could not help it," he told Paul. "The gun went off!"

The victim was a local resident of Mouaou Kombo, the tribe that had welcomed Paul and his men. The village chief accosted Paul. "You say you come here to do no harm and do not kill people," he shouted. "Is not this the dead body of a man?"

Paul was in no position to plead innocence. As the chieftain tried to restore order among his people, Paul ducked into his hut and began to pack up. He feared that the Mouaou villagers might riot at any moment. Paul began throwing journals, bullets, revolvers, watches—everything he could reach—into his bags.

As he hurried, the shouts and curses grew louder. Another woman had just been found dead in her hut, apparently killed by the same stray bullet.

The entire village rose up in a cry for war.

Paul and his men ran.

THEY SPRINTED into the forest under a hail of arrows. One nicked Paul's hand. Another pierced the leg of one of Paul's porters, throwing a limp in the man's stride.

As they fled, every once in a while one of the men would turn around and fire a gunshot at the villagers giving chase. Although each blast seemed to freeze the Mouaou, they continued their pursuit.

Paul's men had a clear advantage in terms of weaponry; the range of their rifles far exceeded that of the bows and arrows. The twisting forest paths protected both sides, because clear shots were practically impossible. As the chase continued, Paul feared the local villagers might know shortcuts through the forest, allowing them to ambush his men.

After about four or five miles, the villagers remained within earshot. Paul called on his men to stop.

"I felt that it was time to make a stand and give them a specimen of our power," he later explained, "for if we allowed them to go on in this way there would be danger of their rousing against us the villagers ahead, and then it would be almost impossible to escape." The porter who'd been wounded in the leg nervously speculated that the locals were using poisoned arrows. It wasn't paranoia: some natives in the interior coated their arrowheads with the toxic extract from a liana when hunting.

The men waited a few moments until they saw the first of their pursuers, and they opened fire. Paul saw a couple of the villagers fall.

They turned to run again, and Paul felt a hot stab of pain in his side.

His leather revolver strap had absorbed most of the arrow's impact, but he'd been hit. Again, his men turned to open fire, while Paul fingered his wound and worried about the possible effects of poison. Unless the firepower convinced their pursuers of their potency, he believed, the villagers might surround them at nightfall and kill them.

They ran to a hill and dug in for their last stand. Paul saw one of the villagers fall and another take a rifle ball in the face, "to all appearance his jaw broken."

This time, the brutal assault worked. When they began to move again, the locals gave up the chase.

PAUL AND his men continued their quick march for several more miles, not entirely convinced that they were safe. Paul estimated they probably traveled twenty miles that day, before they ate and rested. His side hurt, but the wound wasn't serious. The porter who'd been hit in the leg, however, was in great pain, but Paul had no medicine left. In their flight, many of the porters ditched much of their loads, unable to run while carrying them. Paul couldn't blame them. He'd done the same.

For the next two months, they rushed to retrace in reverse the route they'd followed months earlier. In front of them, the smallpox plague had reduced some of the villages to mere ghost towns. Behind them, wars raged.

When the tribes they encountered pressed them for details about their escape, the group "took care to conceal the fact that we were the aggressors," Paul wrote. But their discretion soon wore off. Some of the men began to brag about how many men they had killed.

When Paul reached Quengueza's village, he learned that smallpox had spread through the area shortly after he'd left it. Instead of the thriving settlement he'd known, the area was now barren. Most of the people had moved to another village.

Paul eventually found Quengueza, whose "kingdom" had been obliterated. The old man wanted to leave the coast, the country, the continent. "If I was a young man," he told Paul, "I would go with you to the white man's country. And even old as I am, if your country was not so far off, I would go with you."

Both knew it wasn't going to happen. The Nkomi men who had stuck with him had returned alive, but only Paul would make it to London.

A CAPTAIN bound for England agreed to take him, even though Paul had no way of paying him. He'd lost almost everything during the rush back to the coast.

Somewhere in the jungle, discarded in the bush, were the glass plate photographs, his cameras, the sextants. Of all the equipment he had taken into the forest, only two watches and one barometer made it out.

Just before he boarded the ship, Paul wrote a letter to

Murray, his publisher. It was short, a courtesy to let him know he was alive, though not well.

"You have no idea of the trials I have had to encounter," he wrote. "I have passed through the plague, fire and water."

Throughout the long boat journey to England, a murky lethargy clung to him. It was now almost exactly one year after he arrived in Gabon to begin what Murchison had called the boldest African expedition ever undertaken.

His failure had been nothing short of spectacular.

CHAPTER 40

The Jury of His Peers

The Royal Geographical Society's most anticipated meeting of 1866 was its very first, held on January 8 in Burlington House. Paul, the featured speaker, had first set foot in this building five years before, when he unveiled his gorillas. Now thirty-four, he retained the energy of a young man eager to prove himself. But this time, the tension in the room was of a different character. To Paul, the people filling the lecture hall must have seemed like an enormous jury, eager to examine his evidence and pass judgment. It wasn't simply his expedition that was on trial. It was his dignity.

No gorillas shared the stage. No drawings depicted the slaying of wild beasts. He told no jokes, spun off no ad-libs. He planned to stick closely to his prepared text, which summarized his two-year ordeal, from his canoe spill on the coast to the disastrous escape from Mouaou Kombo. His speech sketched courses of rivers, the customs of tribes, the features of the landscape, his discovery of the Pygmies, the spread of the plague, and, of course, the habits of the gorillas he'd observed.

Transitions were unnecessary, because the thematic glue that held every disparate observation together was the question on everyone's mind: *Was he telling the truth?*

———

PAUL QUICKLY acknowledged that the outcome of the expedition fell far short of its goals. But *all* was not lost. His photographs had vanished, he explained, but the details he provided were delivered with the precision of a meticulous note taker. Instead of judging distances in terms of how many days it took to hike them, he spoke in terms of latitudinal degrees. Instead of "hellish dream creatures," he relied on concrete, observable details when describing the gorillas he encountered in the forest and the others he kept in his encampments. He spoke, in other words, in a language that his audience could trust: that of empirical data.

The information came from his collection of leatherbound notebooks—practically the only things he'd managed to save during his desperate retreat from the jungle. The pages were jammed with statistics. Even on the days that were marred by tragedy and sickness, he had stayed up late in his encampments after the other men had gone to bed, puttering outside by the weak light of a bull's-eye lantern. He'd boil water in a copper kettle, squint at an astronomer's almanac, and fiddle with the sextants. In one notebook, he noted the meridian altitudes of stars to determine his latitude. He charted lunar distances and calculated the altitude of nearly every campsite. He'd also kept a narrative diary with the same obsessive determination to document his every move. On one line he'd describe how the Sword of Orion appeared that night in the sky. Flipping ahead one month, he could find the exact hour of the evening in which he counted thirty-six individual droplets of water on the back of a single leaf. He had notes documenting the length and breadth of the main streets of the villages he visited, the mineral composition of the soil, the distinguishing features of the architecture, and the dress, rites, and even the hairstyles of the people he met. He kept a comparative vocabulary list for each new language he encountered. When stung by an insect, he had noted the species and tried to gauge the intensity and duration of the pain. It was the kind of immersion in the natural world that even photographs might not have captured.

He'd managed to save nearly all those details, and in Burlington House they saved his reputation.

SOME OF the people in the crowd weren't surprised. Even before he had returned to England, several unrelated events had begun to rehabilitate his standing.

Murchison, for example, didn't care that Paul never made it to the Nile. As far as he was concerned, Paul had already proven himself beyond doubt. Early during his stay on the African coast, he had sent back some items with Captain Vardon that aroused quite a bit of interest in the back rooms of the RGS. One was a harp—the same kind of fiber-stringed instrument that had provoked Thomas Malone to mock Paul. In the RGS's official proceedings it was reported that Murchison had "deposited it in the hands of the finest lady harp-player that he was acquainted with, the Duchess of Wellington, who has assured him that musical sounds may be produced from it, though the strings are made from fibres of grass." About the same time, Richard Owen had been thrilled to discover that Paul had sent home a complete preserved specimen of a *Potamogale velox*, the otter-like animal that had been lampooned as the *Mythomys velox* by Gray's army of skeptics. An article in the *Transactions of the Zoological Society* had declared Paul's original description of the animal essentially accurate.

The geographical speculations from the first journey had also been reevaluated during his absence. At the same time that Paul had been winding his disastrous course through Gabon, French exploratory boats had been pushing up the country's Ogowé River. On its first try, the French expedition turned back in the face of native hostilities. But a second attempt gave geographers a better idea of the true course of the river system, and the finding supported many of the geographical conjectures that Paul had included in *Explorations and Adventures*. The German cartographers who earlier had insulted Paul by placing the villages he visited very near the coast redrew their maps, pushing the sites inland.

Finally, in the weeks before Paul's speech, Edwin Dunkin, the head of the Royal Observatory, spent six hours a day poring over the astronomical data in Paul's notebooks. His assistant worked even harder, laboring over the measurements for nine hours a day. After ten days, they still were only half finished. But Dunkin had seen enough to report to the RGS on the same night of Paul's speech that he was "astonished at the multitude and accuracy of M. Du Chaillu's astronomical observations."

John Gray didn't attend the lecture, nor did Charles Waterton; the eighty-two-year-old Squire had died just six months before, after tripping over a bramble while carrying a log outside his mansion. But one of Paul's persistent critics was hard to miss among the hundreds squeezing into the lecture room: Winwood Reade.

At times during the presentation, it seemed as if Paul were addressing his comments directly at the young writer who'd accused him of lying about seeing chest-beating gorillas. "After these opportunities of further observation," Paul said, "I see nothing to retract in the account I have formerly given of the habits of the gorilla." The delivery, given all he'd been through, was defiantly matter-of-fact.

Reade was relatively gracious, considering the bad blood that had flowed between them. He later wrote to the *Times* to praise the explorer he'd previously disparaged. "I will only add that if M. Du Chaillu's forthcoming book is to be as modest and as free from melodramatic gorillas as the paper which he read at Burlington-House," Reade wrote, "he will deserve to be admitted as an authority by *bona fide* scientific men."

CHAPTER 41

Accidental Victories

While Paul had been in Africa, the Darwinian view of natural selection had become increasingly accepted among specialists. Even Owen seemed to soften his criticisms of a theory that was becoming difficult to deny.

In early 1866, Owen published a paper on the anatomy of vertebrates that the *London Review* interpreted as a "significant though partial admission . . . of the truth of the principles of Natural Selection." Owen denied that he'd adopted Darwin's views, responding to the *Review* by saying that his recent paper had simply reiterated ideas about links between species that he'd first proposed in 1850. That being the case, Owen suggested that Darwin—not himself—should be considered the "adopter" in the matter. Darwin, who was still battling illness and had begun working on a book about variations in domesticated plants and animals, was exasperated by what seemed to him an unscrupulous ploy by Owen to turn defeat in the evolution debate into an unearned victory. In a new edition of *On the Origin of Species* published in 1866, Darwin reminded readers that Owen had been arguing against his theory for years. T. H. Huxley was delighted when he read the new edition. "What an unmerciful basting you give 'Our Mutual friend,'" he wrote to Darwin.

As Darwin and his circle of friends became more powerful in Victorian scientific circles, Huxley's debate with Owen over man's relationship to the gorilla—a spat that the press referred to as "the Hippocampus Debate"—continued at a low boil. In 1863, Huxley's *Evidence as to Man's Place in Nature*, the first book devoted wholly to the subject of human evolution, had included an attempt to defeat Owen once and for all when it came to the gorilla's relationship to man. He provided more evidence that versions of the three brain structures Owen had claimed distinguished humans from apes—including the hippocampus minor—could in fact be found in the lesser primates.

As far as Darwin's supporters were concerned, Huxley was the victor. But Owen didn't surrender. Instead, he identified a weak spot in Huxley's argument, modified his own views, and renewed his attack.

OWEN HAD published a monograph titled *Memoir on the Gorilla* in 1865. In it, he tried to clarify his argument—and undercut Huxley's apparent victory—by saying that he never denied that a rudimentary version of the hippocampus minor could be found in apes. Instead, Owen said that the ape version of the hippocampus minor was so different from the human version that it didn't even deserve to be called by the same name. Huxley's "victory" in the hippocampus debate, Owen suggested, was an empty one.

The argument swayed few among Darwin's supporters, many of whom dismissed Owen as a desperate revisionist. But elsewhere in his gorilla memoir, Owen adopted a new strategy of attack that would eventually prove far more effective against Huxley, even if few of his contemporaries recognized it.

In *Evidence as to Man's Place in Nature*, Huxley had sharpened his contention that the differences between men of varying races were as great as those between men and apes. His argument relied on the false and now-disproven assumption that human racial groups could be placed on a straight evolutionary line that progressed from "lower" (blacks) to "higher" (whites). When he wrote that "men differ more widely from one another than they do from the Apes," Huxley was effectively suggesting that blacks were more closely related to gorillas than they were to Europeans. He was wrong, and Owen sensed his vulnerability. Throughout 1865 and 1866, the embattled Owen repeated arguments he'd been making for years, trying to reframe the gorilla debate. He didn't do so because he believed in social equality for blacks; he did it to seize on Huxley's mistake and defend his own persistent belief that humans didn't evolve from gorillas.

Owen's argument hinged on his study of human skulls, including those that Paul provided him with from West Africa. Owen wasn't ready to argue for equality of the races, but he stated, "I have observed individuals of the Negro race in whom the brain was as large as the average one of the Caucasian; and I concur with the great physiologist Heidelberg, who has recorded similar observations, in connecting with such cerebral development the fact that there has been no province of intellectual activity in which individuals of the pure Negro race have not distinguished themselves." The comparatively minuscule cerebral differences between whites and blacks, Owen wrote, marked "the unity of the human family in a striking manner."

When Paul was preparing his second book about his most recent African adventures, Owen once again lent him his full support, agreeing to write an analytic essay to be included as an appendix to Paul's narrative. The essay—a detailed analysis of ninety-three human skulls Paul had collected—praised the explorer as one who'd earned "the respect and gratitude of every genuine lover and student of the science for its own sake." Owen concluded that in all of the African skulls he examined, the essential characteristics that marked them as different from all other species "are as definitely marked as in the skulls of the lightest white races."

Owen hoped that this evidence might prove useful in his argument that men didn't evolve from gorillas. It didn't. Human racial differences and similarities, we now know, are relatively recent developments and practically irrelevant in evolutionary terms. Transitional forms between men and apes are to be found not by looking at racial variations but instead by finding evidence of extinct species in the fossil record— that is, "the missing link."

The overall thrust of Huxley's argument—that scientific evidence suggests men could have evolved from apes—was sound, even though some of his arguments were wrong. The inverse applies to Owen. History has put him on the losing end of the Darwinian debate, but some of the main arguments he used against his enemies now appear correct and honorable. In the Gorilla War, the battles Owen won were those he hadn't intended to fight.

·◦⫞ ⫟◦·

CHAPTER 42

The Explorer

Paul spent much of 1866 in the London suburb of Twickenham at the observatory. The building's owner kept what he called an "open house" for artists and writers, allowing them to work in the building, whose thick walls—reinforced to support the heavy revolving roof—provided silence and sanctuary. Most people used the building as an office, but Paul took full advantage of the offer, moving into the building and adopting it as a semipermanent home. He was given a key to the locked iron bridge that spanned the river and led to the building's door, and he came and went as he pleased.

He often worked alongside Henry Walter Bates to write his book about the second expedition. Bates was a legendary naturalist, having traveled to the Amazon rain forest with Alfred Russel Wallace, the man whose hypothesis concerning how natural selection might influence evolution sparked Darwin to publish his own theory. Bates had stayed in the Amazon for eleven years, shipping back over fourteen thousand species to England, more than eight thousand of which were considered "new." His book of travels, *The Naturalist on the River Amazons*, had been published in 1863 to great acclaim and, more important, to no dispute. With Murray's encouragement, Paul handed his narrative journals over to Bates, who helped shape them into the book *A Journey to Ashango-Land: And Further Penetration into Equatorial Africa.*

Francis Burnand, a celebrated humorist who wrote for *Punch*, often worked in the observatory in a room near Paul's and looked forward to their mutual pipe-smoking breaks for the reliable entertainment they'd provide. Burnand delighted in Paul's inability to sit still, thinking him "a most amusing man, and very excitable." He'd goad Paul with questions about his book or about the geography of equatorial Africa, just to see how he'd respond. Paul "would demonstrate the situation by the aid of tables, chairs, sticks, and anything that came in handy. He would act the stalking of animals, the getting within measurable distance of the gorilla, would show us the almost insuperable difficulties of trapping one of these monster monkeys."

After his successful presentation at the RGS, Paul had regained some of the celebrity that he'd enjoyed in 1861, which helped him indulge what was becoming one of his favorite pastimes: romantic flirtation. He was, according to numerous accounts, an inveterate charmer, the kind of guy who'd rush to open a door for a lady, showering her with smiles and flattery as she passed. "He was a devoted admirer of the fair sex," Burnand wrote of Paul, "and the fairer her fairness the greater was his devotion." Once, when Burnand spotted a beautiful young actress he knew as "Miss Furtado" passing the observatory on a slow-moving pleasure boat, Paul demanded an immediate introduction, and the two men pursued the boat in canoes. They were invited on board. "Within a few moments Du Chaillu had become generally popular, had made friends all round, and had fascinated the attention of La Belle Furtado," who seemed utterly enchanted. Another man, the proprietor of the Adelphi Theatre, also was clearly interested in winning the young actress's attention. He was thrown so off balance by Paul's apparent success, Burnand later reported, that he fell into the river, where he lost his toupee.

WHEN PAUL'S second book was published late in 1866, the reviews, in both England and the United States, were overwhelmingly favorable. Paul was now cast as a victim of unjust suspicion, a genuine explorer who'd patiently overcome the professional jealousy of his rivals.

"Nothing is more characteristic in his book than the admirable temper of its writer," concluded *Chambers's Journal of Popular Literature, Science, and Arts*. "He naturally desired to vindicate himself, and cause his assailants to 'eat dirt.' He has certainly done so; but the dish is cooked daintily, and in the very best French style and taste."

RODERICK MURCHISON and the other eminent men of the Royal Geographical Society affirmed Paul's promotion from adventurer to explorer. They voted to pass an honorary "testimonial" acknowledging his contributions in the fields of geography and natural history, and they awarded him a cash prize to help pay for the loss of his equipment.

"You have all heard how . . . an unlucky accident brought upon him the calamities by which his enterprise was cut short," Murchison told the members of the RGS. "But even as they are, his results are of considerable importance. For he has not only corrected the geography of his former field, but advanced about 150 miles further into the interior than on the previous occasion."

Murchison suggested that the "unshaken gallantry" that Paul had demonstrated during his travels might have been bred into him from birth. "In all respects, I recognize in Monsieur Du Chaillu a true type of the country which gave him birth," Murchison said. "And if France has just right to be proud of him, we also, as well as our kinsmen the Americans, among whom he has also lived, may claim him as our own."

The fact that Murchison apparently didn't know the full story of Paul's heritage wasn't the point. With those words, and the ovation they inspired, Murchison seemed to be offering Paul a tempting possibility: complete acceptance. He finally *belonged* somewhere.

Exactly where that might be, it seemed, was completely up to Paul.

THERE IS a space that exists between truth and falsity, a region where both can commingle and metamorphose into something powerful. It's the place occupied by myths, and it's the territory that Paul explored more thoroughly than any other.

A boy, doomed to a life of inescapable oppression, is cast out into the world. A ragged orphan, he is found washed up beside a river. With the help of a wise teacher and the force of his own determination, he masters skills that will guide him through life. He is tested through a series of far-flung adventures. He slays his beasts. His glory is cruelly snatched away from him, and all seems lost. Facing his most desperate challenge, he summons all the courage and perseverance he has developed over the broad course of his adventures. Finally, he achieves what once seemed ungraspable: freedom.

This is where Paul chose to reside: safe within a mythology of his own making. To live in that realm was to be completely free, but that freedom would have dissolved the moment he committed to any one of the numerous identities he'd been slipping in and out of for years. After his public triumph, Paul could have settled into a role as an authority on equatorial Africa in general or on the region's natural history in particular. If he had chosen to, he could have focused on gorillas, taking advantage of the fact that he had more verifiable expertise concerning this animal than any man alive. But he didn't pursue any of those paths. In fact, he would never attempt to see a gorilla in the wild again. He would never return to Africa.

He could have owned up to his past. But rumors continued to swirl about his ancestry, and Paul never publicly acknowledged the truth. For his entire life, he would keep afloat all the fictions that had made his identity impossible to pin down. He didn't want to be African, or French, or American, or British, or a naturalist, or a showman, or a big-game hunter, or an anthropologist, or a geographical explorer. He clearly wanted to dance between all of them without getting trapped under any one label.

He did not, by any stretch of the imagination, conform to the twenty-first-century ideal of a man who embraces his background and refuses to surrender to the unjust biases against his people; instead, he chose to live in a self-created universe where racial boundaries simply didn't apply to him. Within that mythology, the ultimate objective—the golden ring—was to become a perfectly free and autonomous individual. Freedom of that kind allowed a person to belong everywhere, which is another way of saying that he belonged nowhere in particular.

··◦][◦··

CHAPTER 43

No-Man's-Land

Paul left England shortly after his second book was published, returning to Carmel, New York, where he prepared for a series of lectures that would carry him across a country recovering from the Civil War and Abraham Lincoln's assassination. From Syracuse to Chicago, he told stories that dripped with sensational details—the same kinds of details that had earned him fame, not the kinds that had earned him respect among the learned societies.

On stages around the country, he settled into a character that would cling to him for the rest of his life: *l'ami Paul*, or "Friend Paul," a genial storyteller who could read the crowd with the intuitive reckoning of an old sailor, detecting the strongest currents, riding them for all they were worth. If he heard laughter from the audience when he demonstrated the gorilla's manner of walking, he'd launch into exuberant pantomime, complete with a throaty chorus of sound effects. If people chuckled when he mentioned that many of the tribal kings he'd encountered practiced polygamy, for example, he'd riff on the subject at length, playing it for laughs, indulging the crowd's appetite for exotica with stories of men who had more wives than they could count.

Journalists reflexively described him as a grown-up child, an avuncular sprite. Children loved his act. Immediately after he began his lecture tour, a New York publishing firm asked him to write a book for children based on his real-life adventures.

"They think it will have immense sale here," Paul explained to John Murray, in a letter sent from Carmel in the spring of 1867. "They want it ready for August. I will write it myself and have the English corrected."

That book, called *Stories of the Gorilla Country*, became the first of five popular adventure books for children that he wrote about Africa in the following five years. "Friend Paul" had become a literary brand.

AFTER SETTLING in the United States, Paul rarely returned to England. He kept in touch only sporadically with many of the figures who'd been instrumental in his rise to fame.

Richard Owen still occupied a privileged place in Britain's social and scientific hierarchy, but his influence waned considerably as Darwinism strengthened its foothold. Owen continued to oversee the British Museum of Natural History, and Queen Victoria knighted him. He was even elected to the Club, the most exclusive private society in England, which was founded by Samuel Johnson and limited to only forty members. But Darwin's circle of friends and followers, who now represented a new scientific establishment, continued to distrust him. Years after the gorilla debate had died down, Darwin believed that Owen was still secretly scheming to discredit him and his friends. No longer did Darwin treat his senior colleague with polite deference. Suspecting Owen of writing another anonymous review criticizing his theory in 1872, Darwin wrote of Owen in a letter: "I used to be ashamed of hating him so much, but now I will carefully cherish my hatred & contempt to the last days of my life."

T. H. Huxley became an eloquent advocate for a brand of rational inquiry that ideally remained free from political, religious, and social pressures. Perhaps more than any other scientist of his day, he helped compose the version of history that would define the era for later generations. Because he was the spokesman of the new establishment, Huxley's judgments mattered. After Owen died in 1892, his grandson wrote a laudatory two-volume biography of him. To assess Owen's place in scientific history, an essay was commissioned to serve as an afterword. Huxley wrote it. The sixty-page assessment was measured and gracious, stating that Owen deserved respect despite his work and beliefs regarding transmutation, not because of them. "The reader must bear in mind that, whatever view is taken of Sir Richard Owen's speculations on these subjects," Huxley concluded, "his claims to a high place among those who have made great and permanently valuable contributions to knowledge remain unassailable."

Huxley's final judgment of Paul Du Chaillu carried similar weight among scientists. Although many of Paul's disputed descriptions were eventually supported, he never achieved the unqualified acceptance of the new scientific elite.

Paul and his specimens had been instrumental in driving Huxley to write *Evidence as to Man's Place in Nature*, but the text barely references him. Paul's absence was conspicuous enough to require explanation: Huxley explained that he refrained from quoting Paul's work not because he thought his depictions of the animal in the wild were improbable but because they had been presented in a scientifically unacceptable manner. Paul's melodrama and romance had no place in science, even if many of the discoveries he brought to light were valid.

"It may be truth," Huxley wrote of Paul's work, "but it is not evidence."

THE WRITER J. M. Barrie, famous as the creator of *Peter Pan*, once observed that during his life he'd met three African travelers—Henry Morton Stanley, Joseph Thomson, and Paul—and all three were committed bachelors (Stanley later married, when he was nearly fifty). "One of them says, after returning from years of lonely travel, a man has such a delight in female society that to pick and choose would be invidious," Barrie wrote.

The anonymous comment could have easily come from Paul, who became a close friend of Barrie's and who regularly sprinkled similar asides into his lectures in the United States. As in almost everything else, when it came to personal relationships, Paul liked to portray himself as the ultimate free agent, too committed to an unchained life of travel to commit to any one person or even group of people.

He never married. In the years that followed his expeditions in Africa, he bolstered his reputation as a charming romantic, but aside from numerous but unspecific references in newspaper articles and letters to his "gallantry" and affinity for beautiful women, little or no evidence of romantic relationships exists. More than two decades after Paul's death, the French journalist Michel Vaucaire attempted to unearth evidence of any romantic relationships Paul might have had, but he came up empty. "I have been unable to find the slightest trace of any lasting affair," Vaucaire wrote. "It cannot be said that there was any woman in his life, or any great romance."

Of course, if Paul was hiding a romantic preference for men, it might help explain his lifelong bachelorhood and his discretion concerning intimate relationships. But no evidence of this exists, either.

His personal correspondence strongly hints that the kind of relationship he valued most wasn't romantic at all, but rather the kind of love that binds parents and children. The missionaries John Leighton Wilson and his wife, Jane, who had adopted him as a teenager, provided him with a prototype of filial love that he actively sought to replicate throughout his lifetime. Until the day he died, Paul almost always lacked a fixed address, instead choosing to live as a semipermanent guest in the homes of a rotating cast of prominent couples.

In Philadelphia, he stayed regularly with George W. Childs, a newspaper publisher, and his wife, Emma. In Chicago, the journalist John Anderson and his family kept a special room in their house reserved especially for Paul, who sometimes boarded with them for months at a time. During much of the last decade of his life, Paul lived principally with Charles Daly and his wife, Maria. Judge Daly, as he was known, was a former chief justice of the New York courts, and he had also served as the president of the American Geographical Society, where he had met Paul. Eventually, the two men joined the board of directors of a humanitarian organization called the Philafrican Liberators' League, a group dedicated to securing the freedom of slaves inside Africa, and they became close friends.

Mr. and Mrs. Daly—with Paul in tow—divided their time between a house in Manhattan and their country estate near Sag Harbor. When Maria was dying in 1894, Paul discovered that she had taken pains to guarantee his future.

"She was so motherly to me—*and has remembered me in her will*," he wrote to a friend just after Maria's death. "It was the kinder of her as she had several sisters and a brother. I am writing from the house of Judge Daly and I will stay with him this winter. . . . He is like a father to me."

When Judge Daly died several years later, he left up to twenty-five thousand dollars more in a trust fund for Paul's use, which was a relative fortune before the turn of the century. Paul used the money to travel the world and to continue to explore the no-man's-land between science and myth.

LITTLE COULD have been less likely for a mixed-race child of a French-African island than to develop a strong personal affinity for traditional Nordic culture, but it was oddly natural for someone who felt most at home in the realm of legend.

In the 1870s, Paul—now in his forties—became fascinated with the idea that the English-speaking peoples were descendants not of the Anglo-Saxons but of the Vikings. After spending nearly five years journeying throughout Scandinavia, he wrote a two-volume travel narrative called *The Land of the Midnight Sun*. Eight years later he followed with the eleven-hundred-page *The Viking Age*, which sorted through Norse mythology in an effort to explain the origins of what he considered the "terrible bravery" and "love of conquest" of English-speaking nations.

In the Nordic myths, Paul found historical roots of the kind of heroism that had shaped his worldview from his earliest years. He delved deeply into the stories of characters like Sigurd, the prototypical Viking hero who, according to some of the legends, is born illegitimately, is raised by a foster father who schools him in the art of war, then fulfills his destiny as a hero by slaying a monstrous dragon. In addition to his travel book and mythological opus, Paul wrote his own version of a modern Norse myth, which he published as a novel for boys that he titled *Ivar the Viking: A Romantic History Based upon Authentic Facts of the Third and Fourth Centuries*. The book, which he began writing when he was nearly sixty, includes descriptions of Sigurd (a wise elder in the story) that could have served as snapshots of the author: Sigurd "had traveled far and wide, and seen countries that were unknown to most people; he was short of stature, and had attained the meridian of life; gray hair was beginning to show itself."

Paul's late-to-bloom immersion in Nordic life earned him an energetic following among Scandinavian immigrants in the United States. In 1896, the U.S. senator Knute Nelson lured him to his home state of Minnesota to deliver campaign speeches in support of the Republican presidential nominee, William McKinley, and to capitalize on Paul's "marked weight with the Scandinavians of Minnesota," according to a journalist writing in the *Saint Paul Globe*. The following year, Paul lobbied to land the position as U.S. ambassador to Sweden from McKinley.

Paul lost out in his ambassadorial bid to a veteran U.S. diplomat, but the Scandinavian community in the United States never forgot him. Numerous Scandinavian social clubs from New York to Chicago granted him membership, adopting him as one of their own.

"They are a grand, noble race," he said during one of his New York lectures about the people of Norway and Sweden.

"A good, steady, law-abiding people, and I hope lots of them come here."

Incredibly, the racially ambiguous chameleon who at various times in his life claimed Africa, Europe, and America as his ancestral home ended his life identified more as a Scandinavian than anything else.

IN 1901 a seventy-year-old Paul embarked on his final journey, a trip to Russia that he hoped would result in another travel narrative. Just before he left, he stopped in Washington, D.C., to make some final arrangements with Russian diplomats and the U.S. consulate office. A reporter found him energetically pacing the lobby of the Arlington Hotel, a blunt little plug of a man who couldn't stand still.

"Mr. du Chaillu's books and explorations need no introduction, but his personal appearance does," the journalist wrote. Paul, he observed, still looked remarkably young—boyish, in fact, aside from a white mustache.

"I believe in the power of youth to make one young," Paul explained, "and I avail myself of it at every opportunity. I cultivate young men and women, follow their habits, adapt myself to their customs, and partake of their pleasures. I eschew the gouty, crabbed old ones who pooh-pooh all pleasures and stay at home nursing their aches."

That was why, he said, he was headed to Russia—to undertake another adventure. He planned to travel the entire country, meeting everyone from peasants to the czar, immersing himself in the culture just as he had in Africa and Scandinavia.

The world had changed, entering a period of rapid specialization and industrialization. Adventurers like Paul's old ally Richard Burton, who died in 1890, seemed to belong to an earlier age when large pockets of the world still felt unexplored. Paul seemed reluctant to let go of that time.

"I am restless, curious to see things, but after I will see Russia I will die happy," he said. "I need action. I have been living too high, and it is not good for young men to live too high."

He left for Russia in June 1901 and spent most of his first year in St. Petersburg studying the Russian language, employing two teachers for lessons that lasted six hours a day. A year later, he wrote to a friend, "I hope good health will be granted to me also that I will retain my powers of observations until I have accomplished my work—I want my book on Russia to be one of my best. I suppose you will laugh when you read this and exclaim: Paul does think he is getting old."

At about 10:00 a.m. on April 30, 1903, he was reading a newspaper at breakfast in the restaurant of his hotel when two men heard him speak to no one in particular. "I can't see," he said. "There must be something the matter."

He tried to get up from his chair, but he seemed weak. With help, he staggered to a business office down the hall. Doctors were called, as was a friend who lived in St. Petersburg.

"I am dying," he told his friend before he was taken to his room to lie down. He struggled all afternoon to speak, and when the American ambassador visited his room, Paul seemed to try to remove his right hand from under the bedclothes to greet him, but he couldn't. He was moved to a hospital, where he died at 11:00 p.m.

The American consul placed telegrams to Washington

in an attempt to locate any of Paul's surviving relatives. The response came back: "Have the remains embalmed, placed in a vault, and await instructions." Eventually, the consul's office tracked down some friends who had known him for fifty years, yet none could provide the name of any family member, no matter how distant. Henry Hoyt, who had managed Paul's trust fund as the executor of Judge Daly's estate, in an affidavit reported that despite having a friendship with Paul for twenty years, he "never heard him speak of having any living relative." By default, Hoyt became Paul's executor. But aside from a few scattered papers and personal belongings found in his room in St. Petersburg, Paul had no estate.

He had given away almost everything he owned, to friends and museums, over the course of his lifetime. His "estate" was valued at less than five hundred dollars. A headline attached to a short article printed just after his death in the *New York Times* read, "Du Chaillu Died Poor."

His body arrived in New York on an ocean liner in June. A funeral was held at the Park Presbyterian Church. Representatives from several Scandinavian benevolent societies attended, and the Swedish Glee Club of New York sang a hymn. The Reverend Anson P. Atterbury delivered a short eulogy to the scattering of friends who attended.

"Now that the earthly ending has come to his life, we look back and are surprised to see how strangely varied was that life," Atterbury said.

Paul's gift, the reverend told them, rested in his protean adaptability, the adroitness that allowed him to seek out and thrive in places of dizzying variety, in extreme circumstances, among people of all kinds.

"As an explorer, as lecturer, as author, as social compan-
ion and investigator, he made the world largely his own," the
reverend said as he stood next to the casket. "Known and welcomed on three continents, he had everywhere, yet nowhere,
a home."

⟨⟩

Epilogue

A large upright gravestone, supporting an ornamental granite sphere, sits in a shady grove in the middle of Woodlawn Cemetery in the Bronx. The inscription on the stone reads:

PAUL B. DU CHAILLU
AUTHOR AND AFRICAN
EXPLORER
BORN IN LOUISIANA
IN THE YEAR 1839
DIED AT ST. PETERSBURG
APRIL 16, 1903

The name of the man buried in the vault below the marker is correct, but almost all the details—the place of his birth, the date of his birth, even the date of his death—are wrong. It is a perfectly fitting monument for someone who made certain that a sterile listing of places and dates would utterly fail to capture the truth of his life. But gravestones are never the last word.

IN 1912, Arthur Conan Doyle wrote *The Lost World*, a novel that revitalized the action-adventure genre for the twentieth century, just as the stories of R. M. Ballantyne and H. Rider Haggard had for the nineteenth century. The narrative arc of the novel might have seemed wildly imaginative to readers, but Doyle didn't pluck it out of thin air.

The story follows the explorations and adventures of an expeditionary crew led by Professor George Challenger, whose descriptions of demoniac jungle beasts to the learned societies of London earn him derision as a liar and charlatan. The embattled explorer, true to his name, challenges his most vocal critic to accompany him on a return expedition, where he can prove the veracity of his original descriptions. After a hair-raising escape from pursuing natives, the crew returns to London, where once again Challenger steps onto the platform of the Zoological Institute and earns vindication through the applause of his peers.

More than a decade before the novel was published, Paul Du Chaillu had accompanied Doyle on a tour of Chicago. Later, in 1908, Doyle included a direct reference to Paul in an adventure story he'd set in Gabon that appeared in his *Round the Fire Stories*. But in *The Lost World*, the explorer's name is never mentioned. Doyle's devoted fans have generally agreed that the prototype for Professor Challenger was William Rutherford, a physiologist who taught at the university Doyle attended. It's true, to a point: Rutherford provided Doyle with the superficial details that defined Challenger's character, such as his booming voice and his bushy beard. But Paul Du Chaillu's presence is felt on almost every page.

The same could be said for many of the stories that helped

define the swashbuckling strain of popular culture that permeated the early twentieth century. Jack London, who set the standard for wilderness-based adventures in the twentieth century, said *Explorations and Adventures* was one of the first books he ever read, at age seven. When London died in 1916, both volumes of *The Viking Age* sat on top of his bed stand, occupying the same primary position as they had for five years running. Paul permeates London's stories of the wild, as he does Edgar Rice Burroughs's Tarzan tales. Yet if a reader tries to put a finger on a concrete example of his presence in those works, Paul virtually disappears.

Then there's *King Kong*. Merian C. Cooper's 1933 film set the Hollywood standard for two genres that continue to saturate pop culture: action-adventure and horror stories. Paul's name isn't mentioned in the film, but his influence is unmistakable. Cooper admitted as much in 1965 when he told an interviewer that a turning point in his childhood arrived when his great-uncle handed him an old book written nearly thirty years before he was born. "I made up my mind right then that I wanted to be an explorer," Cooper said. The book was *Explorations and Adventures in Equatorial Africa.*

ON THOSE rare occasions when Paul's name surfaced in the public discourse following his death, it was almost always mentioned in reference to the gorilla. The animal had always sat at the very center of his personal mythology, and it continued to play the central role as his story evolved after his death.

Mountain gorillas were first discovered just six months before Paul died, and this subspecies—less mobile and more approachable than Paul's lowland variety—eventually afforded field scientists better opportunities than ever to amass knowledge of the gorilla's habits in the wild.

Gorillas, it was determined, are not particularly dangerous to humans if they are treated with a sort of deferential respect. As Paul observed, they are fairly strict vegetarians (insects are an occasional exception), not predators. They are capable of great strength and violence, but they are generally harmless if not provoked. It is an incontrovertible fact that human beings pose a far greater threat to gorillas than gorillas do to humans.

As poaching and deforestation in the twentieth century turned gorillas into a vulnerable species, Paul's name became something akin to a mild slur. He symbolized human arrogance, the man who invented the damaging stereotypes that had turned people against gorillas, planting false impressions that now threatened the very survival of the species. His life story was abbreviated into an archetype that, given his personal history, was full of irony: Paul was the hubristic colonial interloper who abused a foreign continent and every living thing on it.

Paul's exaggerations and self-serving stories deserved criticism, but they didn't change the fact that even a century after his travel, he'd still given the world more accurate information about lowland gorillas than anyone else. Twentieth-century researchers definitively established that gorillas really do stand on two legs, beat their chests, and charge humans with terrifying swiftness if they are threatened, just as he described. Paul didn't know that such displays are usually meant to scare threats away, not to inflict actual physical harm. If he had simply stood still and assumed a nonthreatening posture, as he did the time he stumbled upon a gorilla without his gun, those charging gorillas almost certainly would have pulled up well short of him, veering off to the side instead of attacking. But achieving that sort of composure— simply to stand still when staring down an animal that at the time was wholly unknown—would have been anything but simple. Paul had no way of knowing that the gorillas likely wouldn't have hurt him had he lain low, and it's unreasonable to have expected him to intuit it. His errors and exaggerations can't be ignored, but they can be understood. He was astonished by the mysterious animals he encountered in the forest because they were, in the eyes of one who viewed the animals without the benefit of any frame of reference, truly astonishing. When we realize this, the fact that a young man with no formal training got so much *correct* becomes more noteworthy than his shortcomings.

George Schaller, a field biologist who in 1960 published the first thorough study of mountain gorillas in the wild, lamented the fact that Paul's imperfections had caused his work to be forgotten. "This was unfortunate, for he was basically a competent and reliable observer," Schaller wrote.

Now, in the twenty-first century, Paul is undergoing yet another reevaluation. A recent academic collaboration between the University of Lyon and Omar Bongo University in Gabon united French and Gabonese historians, ethnologists, primatologists, linguists, and geographers who retraced parts of Paul's route. Together they produced a series of scholarly articles that aimed to assess the explorer's impact on the country. Those articles were collected and published by the French National Center for Scientific Research as *Cœur d'Afrique: Gorilles, cannibales et Pygmées dans le Gabon de Paul Du Chaillu* (Heart of Africa: Gorillas, Cannibals and Pygmies in the Gabon of Paul Du Chaillu).

Taken as a whole, the project represents an overwhelmingly positive reassessment of his place in the history of equatorial Africa. The articles collectively conclude that the information Paul provided about the country's natural history, languages, and ethnology was mostly accurate and surprisingly varied. Given the dearth of written sources that predated Paul's, his work has become an invaluable starting point for serious scholarly research of Gabon's history.

Paul had correctly discovered links between the roots of different native languages, had accurately described tribal customs that had never been observed before, and helped modern scholars imagine life in a time and place that lacked a written history. Jean-Marie Hombert, a French linguist, and Louis Perrois, a French ethnologist, wrote that Paul's investigations of the country's oral histories, languages, ethnological makeup, and natural habitat "allow us to perceive through him a historical, cultural, and natural landscape that in the end is quite different—and certainly more complex and accurate—than we would have ever dared to imagine before."

The concluding essay in that collection was titled "The Posthumous Revenge of a Forgotten Explorer."

REVENGE CAN take root in unexpected quarters and require generations to reach fruition. It can also drip with an irony that can be appreciated only in hindsight.

The resuscitation of Paul's reputation within Gabon itself can be traced back to the man who'd been his most bitter enemy there: R. B. N. Walker, the English trader who in 1861 accused Paul of inventing his adventures and publicly cast aspersions on his parentage and racial background.

Among the episodes that Walker accused Paul of inventing was his supposed climbing of Mount Andele in the Eshira territory, the farthest point inland that he claimed to visit during his first expedition. Even after Paul's second expedition vindicated some of his descriptions from his first book, some geographers remained skeptical as to whether he had actually reached the Eshira territory, or instead—as Walker argued—had simply relied on descriptions from natives to describe it.

Walker lived in Gabon for years after his attacks on Paul. He died in England in 1901, passing his meager assets to a son named Harry, who lived in Surrey. His will made no mention of the other children he had left behind in Gabon—at least a dozen of them, the mixed-race children born to various Mpongwe women.

One of his offspring, a boy named André Raponda Walker, was born in 1871. Educated by Catholic missionaries, he grew into one of Gabon's most esteemed twentieth-century scholars. Eventually, he was awarded the title monsignor by Pope John XXIII.

While researching a 1960 history book, Raponda Walker perused the writings of the Catholic priests who founded the first monastery in the interior of Gabon. The priests' journals indicated that they explored the region around the monastery and climbed the highest mountain they saw. The Eshira natives told the priests that the mountain was called Mukongu-Polu.

In the local language, it means "Paul's Mountain." In his book, Raponda Walker wrote that on the top of the peak, the priests found a large boulder. A name had been etched into the rock many years before: "Paul Du Chaillu."

Acknowledgments

Lots of people contributed to this project, and I'm grateful to all of them for their assistance and support. In Africa, the staff members of the *Projet Gorille Fernan-Vaz* (Fernan-Vaz Gorilla Project) generously provided me an education in lowland gorilla behavior and their habitat in Gabon. Nick Bachand, a field veterinarian and the project's coordinator, tirelessly answered my questions and introduced me to several other researchers who enriched my research and travels. Herman Loundou Ibouanga Landry, Mbembe Jean-Louis, Mbini Joseph-Banu, and Taty Makoty Joris Pierre offered insights into Gabon's forest folklore and local perceptions of gorillas. Josephine Head, of the Max Planck Institute for Evolutionary Anthropology, graciously welcomed my surprise visit to her campsite, then cheerfully shared her knowledge of lowland gorillas in the wild. Without the help and hospitality of Africa's Eden tour operators in Gabon, I don't think I would have found any of them.

I'm also indebted to Christopher E. Cosans, author of *Owen's Ape and Darwin's Bulldog: Beyond Darwinism and Creationism,* for commenting on parts of the manuscript and helping me understand some of the technical points of the debate between Owen and Huxley. Owen biographer Nico-laas Rupke (*Richard Owen: Victorian Naturalist*) also generously reviewed parts of the manuscript and offered valuable advice. French linguist Jean-Marie Hombert provided me a copy of *Cœur d'Afrique*—a book of scholarly essays about Du Chaillu that he compiled with the support of the French National Center for Scientific Research and Omar Bongo University in Gabon. Others who generously donated their time and talents include Adam Lifshey, Eric Eason, and Calvin Sloan.

Larry Weissman and Sascha Alper aren't only the best agents I can imagine, but they're also indispensable advisors, discerning readers, and wonderful people. I've been very for-tunate to work with some exceptional professionals at Dou-bleday, including Jackeline Montalvo, Phyllis Grann, Michael Windsor, Bette Alexander, Lorraine Hyland, Ingrid Sterner, Maria Carella, Todd Doughty, Andrew Sharetts, and Judy Jacoby. My editor, Melissa Danaczko, is an author's dream—creative, smart, and inspiringly enthusiastic—and I lucked out big-time when she decided to take on this book and shep-herd it toward publication.

Finally, my biggest thanks go to the three loves of my life: Mei-Ling, who has given more to this book than anyone else; Violet, who was born somewhere around Chapter 7; and Sofia, who really wanted me to write a book about elephants.

A Note on the Sources

Paul Du Chaillu's celebrity had largely faded by the time of his death, which meant that few took interest in systematically collecting the papers and letters he left behind. But some of his writings surfaced in the archives of contemporaries whose legacies were considered more secure than his.

The Royal Geographical Society provided me with the correspondence that Du Chaillu exchanged with its members in the 1860s. Du Chaillu's letters to his publisher are housed in the John Murray Archive at the National Library of Scotland, and these proved especially helpful in reconstructing Du Chaillu's activities in 1861 and 1862. The Academy of Natural Sciences in Philadelphia provided letters Du Chaillu wrote to its members in the 1850s, as well as the academy's internal notes regarding its strained relations with him.

Some of the papers that Du Chaillu left behind upon his death are stored in the archives of Charles and Maria Daly, which I reviewed in the Manuscripts and Archives Division of the New York Public Library. At the Wisconsin Historical Society archives in Madison, I read through the diaries and letters of the African missionary William Walker, which included references to Du Chaillu and letters regarding him, Winwood Reade, and Richard Burton.

Margaret Clifton at the Library of Congress showed incredible patience in trying to help me track down elusive newspaper articles, and I'm similarly grateful to Susan Duncan and Cinda Pippenger at the University of Illinois library. Joy Wheeler at the RGS helped me attempt to re-create the atmosphere of the organization's meetings in the 1860s, allowing me to pin down the locations of specific RGS gatherings and then providing historical photographs and architectural details to aid my descriptions.

Additionally, I made frequent use of the Darwin Correspondence Project (www.darwinproject.ac.uk), which has electronically indexed the naturalist's letters to and from many of the scientists mentioned in this book. Several other correspondence collections also provided small details in the narrative reconstruction, and I have referenced these individually in the Notes.

Newspaper coverage of Du Chaillu and his gorillas was invaluable, and of the newspapers I consulted via databases and libraries, the archives of the *Times* of London and the *Athenaeum* proved particularly helpful.

Because of Du Chaillu's reticence and deceptions concerning the circumstances of his birth and early life, sources often contradict one another. In trying to sort them out, I am indebted to Henry Bucher Jr., whose digging through records in Gabon and France provided the first solid evidence of Du Chaillu's early history. The Gabonese historian Annie Merlet's research into the life of Paul's father was also invaluable.

Throughout the book, dialogue that is set off by quotation marks reflects the exact transcriptions of the materials referenced in the Notes, with the following exceptions. Until the mid-twentieth century, Gabon was often referred to in print as "the Gaboon" or, less commonly, "the Gabun." For consistency and clarity, I have standardized the spelling in most references throughout the book, including inside direct quotations. Also, nineteenth-century newspapers sometimes transcribed the minutes of public events and printed the words of the speakers using past-tense constructions—even though the speakers had used the present tense when addressing their audiences. In these cases I restored the present tense when directly quoting the speaker; I cite all instances of such changes in the Notes.

Notes

PROLOGUE

xvii *The only sounds seemed to come*: The impressions, texture, and specific details surrounding Du Chaillu's first glimpse of gorillas come from his own descriptions of his initial encounter. These sources include Du Chaillu, *Explorations and Adventures*; Du Chaillu, *Stories of the Gorilla Country*; and an unsigned dispatch titled "Rare Animals from Africa," authored by Du Chaillu and reprinted in the *Massachusetts Spy*, July 6, 1859.

xix *"like men running for their lives"*: Du Chaillu, *Explorations and Adventures*, 60.

CHAPTER 1. DESTINY

3 *Late in 1846, near the end of the rainy season*: Gabon generally experiences two rainy seasons each year, the second of which runs between September and December.

3 *But these men had been exploring*: Descriptions of trade customs and native traders come principally from Wilson, *Western Africa*; Du Chaillu, *Explorations and Adventures*; and Aicardi de Saint-Paul, *Gabon*.

3 *the traders carried something extraordinary*: Details of Wilson's acquisition of the skull from native traders come from DuBose, *Memoirs*; Wilson, *Western Africa*; Leonard G. Wilson, "The Gorilla and the Question of Human Origins: The Brain Controversy," *Journal of the History of Medicine and Allied Sciences* 51, no. 2 (1996); Thomas Savage and Jeffries Wyman, "Notice of the External Characters and Habits of Troglodytes

Gorilla, a New Species of Orang from the Gaboon River," *Boston Journal of Natural History* 5, no. 4 (Dec. 1847); "Wild Men of the Woods," *Household Monthly* 1, no. 6 (March 1859); and Du Chaillu, *Explorations and Adventures.*

4 *He could spend hours marveling*: DuBose, *Memoirs*, 175–76.

4 *At first glance, that calcified mask*: The weight, measurements, and physical details of the skull were provided by Savage and Wyman, "Notice"; Wilson, *Western Africa*; and DuBose, *Memoirs.*

5 *The natives called it a* njena: The Mpongwe term has been transcribed by visitors to Gabon using numerous spellings, including *ngena, d'jina, engina*, and other variations; I use *njena* throughout for consistency.

6 *"pliancy of character"*: Wilson, *Western Africa*, 31.

6 *"descendants of Ham"*: DuBose, *Memoirs*, 194.

6 *"That this people should have been preserved"*: Ibid., 199.

7 *Wilson one day in 1848 spotted*: Dates and details of Du Chaillu's arrival at Baraka come from William Walker, diaries, Wisconsin Historical Society, Madison; DuBose, *Memoirs*; and Helen Evertson Smith, "Reminiscences of Paul Belloni Du Chaillu," *Independent* 55 (1903): 1147.

8 *a messy process that Wilson and Jane*: French dealings with King Glass and descriptions of Baraka in the 1840s and 1850s come from DuBose, *Memoirs*; Wilson, *Western Africa*; Aicardi de Saint-Paul, *Gabon*; Rich, *A Workman Is Worthy of His Meat*; Patterson, *Northern Gabon Coast to 1875*; Meyer, *Farther Frontier*; and West, *Congo.*

8 *"It is doubtful"*: DuBose, *Memoirs*, 166. In the memoir, Jane was directly quoted, though her words were rendered in the past tense ("It was doubtful"); I changed it to present tense for clarity.

9 *Wilson realized that he in fact had met Paul's father*: Walker, diaries; and Henry Bucher Jr., "Canonization by Repetition: Paul Du Chaillu in Historiography," *Revue Française d'Histoire d'Outre-Mer* 66 (1979): 15–32.

9 *The trader, named Charles-Alexis*: Du Chaillu's father's name appears as Claude-Alexis in certain references, but I have used Charles throughout.

10 *"delicate little attentions"*: DuBose, *Memoirs*, 82.

10 *he started calling them "Father" and "Mother"*: Ibid., 153.

11 *In the boy's eyes, Wilson was a miracle*: After Wilson's death, Du

Chaillu once was heard snapping at someone who suggested that Wilson was "narrow" by exclaiming, "He was as broad as this broad earth!!" (ibid., 147).

12 *"The dog experienced no injury"*: Wilson, *Western Africa*, 368.

CHAPTER 2. A NEW OBSESSION

13 *Caroline Owen tried very hard*: Descriptions of the Owen household and excerpts from Caroline's diaries come from Owen, *Life*; and Yanni, *Nature's Museums*.

13 *"The presence of a portion"*: Owen, *Life*, 1:296.

15 *"himself smelt like a specimen preserved"*: Ibid., 233.

16 *"I have found the existence of an animal"*: Savage to Owen, April 24, 1847, reprinted in *Transactions of the Zoological Society of London,* no. 3 (1849): 389.

16 *"Great uncertainty however attends"*: Ibid.

17 *A naturalist in Bristol whom Owen knew*: Details of Owen's examinations and approach come from Richard Owen, "On a New Species of Chimpanzee," *Proceedings of the Zoological Society of London* 16 (1848): 27–35; Richard Owen, *Memoir on the Gorilla* (London: Taylor and Francis, 1865); and Richard Coniff, "The Missionary and the Gorilla," *Yale Alumni Magazine*, Sept.–Oct. 2008.

CHAPTER 3. HANNO'S WAKE

18 *The name "gorilla" was plucked*: Although some sources give Wilson, who was familiar with Hanno's story, credit for naming the gorilla and passing it to Savage, it was Wyman who chose the name. A concise overview of the naming process is included in Jordan, *Leading American Men of Science*.

18 *"sound of pipes, cymbals, drums"*: Hanno and Falconer, *Voyage of Hanno.*

19 *"They goe many together and kill"*: Purchas, *Hakluytus Posthumus*, 6:398.

19 *"Its paw was said to be"*: Savage and Wyman, "Notice."

20 *"It is almost impossible to give a correct idea"*: Wilson, *Western Africa*, 367.

20 *"It is said they will wrest"*: Ibid.

21 *"This act, unheard of before"*: Savage and Wyman, "Notice."

22 *Wilson sent some letters*: Bucher, "Canonization by Repetition."

CHAPTER 4. DRAWING LINES

23 *"Mr. Darwin was here"*: Owen, *Life*, 1:292.

24 *"ends are obtained"*: Richard Owen, *On the Archetype and Homologies of the Vertebrate Skeleton* (London: John van Voorst, 1848).

24 *Darwin pleaded with him*: Owen, *Life*, 1:209.

24 *The day of creation, it stated*: Larson, *Evolution*.

25 *The scientific evidence for questioning the Bible's time line*: For the brief overview of early evolutionary science, two particularly helpful books were ibid., and Bowler, *Evolution*.

27 *"a real man in armour is required"*: Owen, *Life*, 1:254.

27 *His ideas about archetypes*: For a more detailed study of Owen's views on evolution, see Cosans, *Owen's Ape and Darwin's Bulldog*.

CHAPTER 5. AMERICAN DREAMS

28 *Carmel, nestled between rolling hills*: Descriptions of nineteenth-century Carmel come from Blake, *History of Putnam County, N.Y.*; George Carroll Whipple III, *Carmel* (Charleston, S.C.: Arcadia, 2007). The historical population figures come from J. H. French, *Gazetteer of the State of New York* (Bowie, Md.: Heritage Books, 2007).

29 *he spoke it with a soupy Parisian accent*: Du Chaillu's accent was the subject of much discussion and gentle chiding over the years; Evertson Smith, for example, wrote of him speaking of "My country, dese United States."

29 *He told his students that he despised*: Evertson Smith, "Reminiscences."

29 *he visited the Putnam County Courthouse*: Ibid., and Bucher, "Canonization by Repetition."

30 *"Mr. Du Chaillu's diminutive size"*: Evertson Smith, "Reminiscences."

31 *The editors agreed to publish*: Vaucaire, *Gorilla Hunter.*

32 *Thus, on October 16, 1855, Cassin stood*: "Proceedings of the Academy of Natural Sciences of Philadelphia," reprinted in *Littell's Living Age*, 3rd ser., 14 (July–Sept. 1861).

CHAPTER 6. TO SLIDE INTO BRUTISH IMMORALITY

33 *Three were human skulls*: Richard Owen, "Notices of the Proceedings of the Royal Institution," 1855.

33 *Chimpanzees had been known*: Ibid.

33 *orangutans had first been described*: Beeckman, *Voyage to and from the Island of Borneo.*

33 *"It is not pleasing to me that I must place"*: Linnaeus to Johann Georg Gmelin, Feb. 25, 1747. The Linnaean Correspondence, available online at http://linnaeus.c18.net/Letters/display_txt.php?id_letter=L0783.

35 *"The present age may be more knowing"*: Owen, "Notices."

35 *"And of a truth, vile epicurism"*: More and Descartes, *Collection of Several Philosophical Writings of Dr. Henry More.*

CHAPTER 7. AN AWKWARD HOMECOMING

36 *The captain's charts showed*: The description of the journey to Africa is from Du Chaillu, *King Mombo.* Additional descriptions of the surf around the estuary in the nineteenth century can be found in Burton, *Two Trips to Gorilla Land.*

37 *"For five days"*: Du Chaillu, *King Mombo*, 23.

37 *"I had enough powder"*: Ibid.

38 *Shortly before Paul arrived*: DuBose, *Memoirs.*

38 *But the old man's presence*: Du Chaillu, *Explorations and Adventures.*

38 *Charles-Alexis Du Chaillu was interred*: Burton, *Two Trips to Gorilla Land.*

39 *"Their disappointment was great"*: Du Chaillu, *Explorations and Adventures*, 3.

40 *"It is really tanned a very dark brown"*: Ibid., 61.

41 *Thousands of yards of cloth bundled*: Good overviews of the supplies carried into Africa by nineteenth-century explorers can be found in Burton, *Lake Regions of Central Africa*; Stanley, *How I Found Livingstone*; and Jeal, *Stanley.*

43 *"But the more he called 'stop'"*: Du Chaillu, *Explorations and Adventures*, 53.

44 *"A good day's work"*: Ibid.

46 *"Dayoko thought my project impossible"*: Ibid., 60.

46 *Britain sent four major expeditions*: Jeal, *Stanley.*

46 *In Zanzibar, they were told*: Rice, *Captain Sir Richard Francis Burton.*

47 *"dig violently away at my tympanum"*: Speke's quotation was taken from an 1859 issue of *Blackwood's* magazine and reprinted in Burton, *Lake Regions of Central Africa.*

48 *"No one would tell me how he was to be killed"*: Du Chaillu, *Explorations and Adventures*, 39.

49 *"Oh yes," one of the men replied*: Ibid., 40.

CHAPTER 8. "UNFRIENDS"

50 *John Edward Gray couldn't stand*: Gray's animosity toward Owen is explored in Joel Mandelstam, "Du Chaillu's Stuffed Gorillas and the Savants from the British Museum," *Notes and Records of the Royal Society of London* 48, no. 2 (July 1994); and Rupke, *Richard Owen*.

50 *As a young man, he had wanted*: Gunther, *Century of Zoology at the British Museum*; and G. S. Boulger, "John Edward Gray," in *Dictionary of National Biography*, vol. 23 (London: Smith, Elder, 1900).

51 *product of its creator's "prurient mind"*: Gunther, *Century of Zoology at the British Museum*.

51 *"One can easily understand that the circumstance"*: "Biographical Notice of the Late Dr. J. E. Gray," *Annals and Magazine of Natural History* 15 (1875).

52 *Darwin was told that Gray*: Darwin to Gray, Aug. 29, 1848.

52 *"I felt anxious to know"*: Ibid.

52 *Gray backed down and continued*: Gray to Darwin, Aug. 26, 1848.

52 *Surrounded by lush gardens*: Descriptions of Owen's home at Sheen Lodge are from Owen, *Life*, vol. 2.

52 *But in 1856, Owen was given another gift*: Rupke, *Richard Owen*.

53 *"In a year or two, the total result"*: Huxley to William Macleay, Nov. 9, 1851.

CHAPTER 9. FEVER DREAMS

54 *Instead, almost everyone subscribed to a "miasma theory"*: For a good overview of the theory and its demise, see Johnson, *Ghost Map*.

55 *"all the authorities on this subject"*: Reade, *Savage Africa*.

55 *In the mid-nineteenth century, 2 grains of quinine*: "The Use of Quinine in Malarious Districts," *Boston Medical and Surgical Journal*, Oct. 1, 1863.

55 *"And when the system becomes accustomed"*: Du Chaillu, *Explorations and Adventures*, 323.

56 *In recent years, medical researchers have determined*: Combiz Khozoie, Richard J. Pleass, and Simon V. Avery, "The Antimalarial Drug

Quinine Disrupts Tat2p-Mediated Tryptophan Transport and Causes Tryptophan Starvation," *Journal of Biological Chemistry*, no. 284 (2009).

56 *distorted states of perception that he labeled "ecstasis"*: Fabian, *Out of Our Minds*.

57 *"Today {August 20} I sent back Dayoko's men"*: Du Chaillu, *Explorations and Adventures*, 48. In this case in his book, Du Chaillu transcribed his journals in the present tense.

59 *"Peering into the darkness"*: Ibid., 112.

60 *One of the men told Paul a story*: In Gabon in 2010, I spoke with numerous Gabonese residents and hunters in the Fernan-Vaz region who told me stories about relatives or acquaintances who, they said, had been abducted by gorillas. Some admitted that the stories were likely legends, but others swore to me that they were true, illustrating that the same forest folklore that Du Chaillu recorded still thrives today.

60 *"Yes," one of the Mbondemo men told Paul*: Du Chaillu, *Explorations and Adventures*, 61.

CHAPTER 10. BETWEEN MEN AND APES

62 *"It is astonishing with what an intense feeling"*: Huxley to William Macleay, Dec. 13, 1851.

63 *"his destiny as the supreme master of this earth"*: Richard Owen, "On the Characters, Principles of Division, and Primary Groups of the Class Mammalia," *Journal of the Proceedings of the Linnean Society* 1–2 (1857).

63 *a vindicator of "the dignity of the human race"*: "A Monkey, Not a Man," *London Lancet* 1 (1859).

63 *"As these statements did not agree"*: T. H. Huxley, *Man's Place in Nature*.

CHAPTER 11. MAPS AND LEGENDS

64 *"The dry season is delightful in Africa"*: Another present-tense journal entry transcribed in Du Chaillu, *Explorations and Adventures*, 37.

65 *Before someone like John James Audubon*: Rhodes, *John James Audubon*.

66 *He had catalogued and named*: "John Cassin," *Bulletin of the Essex Institute* 1 (1869).

66 *had already shipped him more than one thousand specimens*: Secretary's records in the Archives of the Academy of Natural Sciences of Philadelphia.

67 *He had never heard a sound so unsettling*: In 2010, I visited Josephine

Head, who runs a gorilla research station in Gabon. Her project, which aims to habituate a group of lowland gorillas to human presence for scientific observation, requires her to passively observe charging silverback males. "It's a *huge* roar. Absolutely terrifying! I've heard it hundreds of times now, and every single time I hear it, I shake. A gorilla will hide silently, then suddenly out of nowhere you get this massive roar. Everything about it is designed to intimidate, and it works." She added: "You have to just stand there and try not to move, because if you run away, he will chase you and grab you and bite you. So you have to *withstand.*"

68 *appeared to stand almost six feet tall*: Du Chaillu was struck by its height and, after measuring it, first reported that it was "two inches short of six foot," and in another reference he listed its height as five feet eight; the latter is probably more reliable, because male lowland gorillas rarely exceed that height.

68 *"Luckily, one of the fellows shot a deer"*: Du Chaillu, *Explorations and Adventures*, 71.

69 *cheap, African-made flintlock muskets*: Information about the character and quality of native muskets can be found in Reade, *Savage Africa*; Du Chaillu, *Explorations and Adventures*; and Gavin White, "Firearms in Africa," *Journal of African History* 12, no. 2 (1971).

70 *In 2001, scientists split gorillas*: Caldecott and Miles, *World Atlas of Great Apes and Their Conservation*.

70 *But they are also the least studied in the wild*: Much of the information about lowland gorillas comes from interviews with Josephine Head, a lowland gorilla specialist based in Gabon with the Max Planck Institute for Evolutionary Anthropology. I also consulted several books concerning gorilla behavior, and the most helpful were Schaller, *Year of the Gorilla*; Weber and Vedder, *In the Kingdom of Gorillas*; and Fossey, *Gorillas in the Mist*.

72 *"a wicked man turned into a gorilla"*: Du Chaillu, *Explorations and Adventures,* 298.

CHAPTER 12. A LION IN LONDON

74 *"the distinguished traveler we have this day"*: "Farewell Livingstone Festival," *Proceedings of the Royal Geographical Society* 2–3 (1858); and Owen, *Life*, vol. 2.

74 *"that higher wisdom which is not of this world"*: "Farewell Livingstone Festival."

75 *"That is the will of David Livingstone"*: Blaikie, *Life of David Livingstone*.

75 *Murchison had personally micromanaged*: Overview of Murchison's background before meeting Du Chaillu comes from Stafford, *Scientist of Empire*; Geikie, *Life of Sir Roderick Murchison*; and Mill, *Record of the Royal Geographical Society*.

CHAPTER 13. THE MAN-EATERS

78 *Today they're the dominant ethnic group*: Hombert and Perrois, *Cœur d'Afrique*. Although Du Chaillu and most other nineteenth-century writers referred to them as "Fan," I've complied with the modern spelling, "Fang."

78 *It started with Herodotus*: Herodotus, *Histories*.

79 *In his journal on November 23, 1492*: *Journal of Christopher Columbus*. A good overview of Columbus's views of cannibals can be found in Peter Hulme, "Columbus and the Cannibals," in *The Post-Colonial Studies Reader*, ed. Bill Ashcroft, Gareth Griffiths, and Helen Tiffin (New York: Routledge, 1995).

79 *described by everyone from Captain Cook to Herman Melville*: A good overview of cannibal reports can be found in David F. Salisbury, "Brief History of Cannibal Controversies," *Exploration: The Online Research Journal of Vanderbilt University*, Aug. 15, 2001.

79 *Queen Isabella of Spain issued a decree*: Ibid.

81 *"Mbene is in great glee"*: Du Chaillu, *Explorations and Adventures*, 76.

82 *"Today, several hundred Fang from the surrounding"*: Ibid., 80.

82 *"Why do you come from nobody knows"*: Ibid., 491.

CHAPTER 14. D.O.A.

83 *Owen and a taxidermist named Abraham Bartlett*: "Wild Men of the Woods."

84 *"Whether we shall ever be treated"*: "A Monkey, Not a Man."

CHAPTER 15. SPIRIT OF THE DAMNED

85 *"Looking upstream almost any time"*: Du Chaillu, *Explorations and Adventures*, 196. Hombert and Perrois, in *Cœur d'Afrique*, located the

present-day site of Du Chaillu's "village" and provide a corroborative history of the slave trade in the region.

87 *"They seemed terrified out of their senses"*: Du Chaillu, *Explorations and Adventures*, 145.

87 *"The place had been used"*: Ibid., 180.

88 *Most of the coastal tribes*: Ibid., and Wilson, *Western Africa*.

88 *"Even in this rude Cape Lopez country"*: Du Chaillu, *Explorations and Adventures*, 181.

89 *"The young one, hearing the noise"*: Ibid., 207.

89 *"All the hardships I had endured"*: Ibid., 205.

90 *"He sat in his corner"*: Ibid.

91 *"As soon as he saw his mother"*: Ibid., 244.

91 *"Though there are sufficient points of diversity"*: Ibid., 277.

92 *"I imagined this repulsive aspect originates"*: Darwin, *Voyage of the* Beagle.

92 *"It was as though I had killed"*: Du Chaillu, *Explorations and Adventures*, 434.

CHAPTER 16. ORIGINS

93 *"massive popular success"*: Katherine Haddon, "Darwin at 200: Modest Father of Biology," *Cosmos Magazine*, Feb. 11, 2009.

93 *John Murray originally printed*: A good overview of the sales history of *On the Origin of Species* can be found in David B. Williams, "Benchmarks: *On the Origin of Species* Published," *Earth*, Nov. 23, 2009.

94 *Almost half of British adults of marrying age*: Woods, *Demography of Victorian England and Wales*.

94 *"probably all the organic beings"*: Darwin, *On the Origin of Species*.

94 *"I {should} be a dolt not to value"*: Darwin to Owen, Dec. 13, 1859.

95 *Owen asked how, if all life-forms*: "Darwin on the Origin of Species," *Edinburgh Review*, April 1860, 487–532.

96 *"Mr. Darwin abhors mere speculation"*: "Darwin on the Origin of Species," *Times* (London), Dec. 26, 1859.

96 *"I am prepared to go to the Stake"*: Huxley to Darwin, Nov. 23, 1859.

96 *"I am sharpening my claws"*: Ibid.

96 *When word spread among London's scientific community*: The knowledge of Owen's authorship is evident in letters exchanged in 1860 between Darwin, Huxley, and Asa Gray.

CHAPTER 17. IN THE CITY OF WONDERS

97 *a city pulled in two different directions*: Spann, *Gotham at War.*

97 *But at seven o'clock on the evening*: Descriptions of the event and press clippings come from *Official Report of the Great Union Meeting Held at the Academy of Music* (New York: Davies & Kent, 1859).

98 *Newspapers covering the event*: Ibid.

98 *"Hundreds and thousands are among us"*: Ibid.

99 *"the condition the Negro is assigned"*: Ibid.

99 *"Experience has shown that his class"*: Ibid.

99 *In the neighborhood of Five Points*: Several sources contributed to the description of Five Points in the nineteenth century, including Asbury, *Gangs of New York*; Harris, *In the Shadow of Slavery*; and Dickens, *American Notes.*

99 *"Debauchery has made the very houses"*: Dickens, *American Notes.*

100 *"To give you a correct and critical description"*: Bobo, *Glimpses of New York City.*

101 *"for a shoe clerk out of a job"*: "Man Who Alone Captured Brazilian Navy Is Here," *New York Times*, Oct. 20, 1912.

102 *members of the academy were acting*: Based on letters exchanged between Du Chaillu and the Academy of Natural Sciences of Philadelphia in Dec. 1859.

102 *"possesses peculiar advantages as an explorer"*: John Cassin report to the Academy of Natural Sciences of Philadelphia, Oct. 6, 1855.

103 *"Under the circumstances of the case"*: Du Chaillu to the Academy of Natural Sciences of Philadelphia, Jan. 31, 1860.

104 *"French traveler who had advanced"*: "Notices," *New-York Daily Tribune*, Jan. 6, 1860.

104 *"They died very easily"*: Ibid.

104 *Wyman in turn invited Paul*: Vaucaire, *Gorilla Hunter*; and "An Editor Taking Notes Among Celebrities and Others," *New York Times*, March 17, 1912.

105 *"DU CHAILLU'S AFRICAN COLLECTION"*: Advertisement in *New-York Daily Tribune*, Feb. 14, 1860.

105 *It was marble fronted, with Italianate*: Historical building details from New York City Landmarks Preservation Commission, *Guide to New York City Landmarks.*

105 *"Hideous monsters with unearthly names"*: "The Gorilla," *New York Post*, March 29, 1860.

106 *His ads during the first weeks of 1860*: Based on recurring advertisements for Barnum's American Museum in the *New-York Daily Tribune* in Jan. and Feb. 1860.

107 *"If such a being as Zoe ever existed"*: Adams, *E Pluribus Barnum*.

107 *"warmly towards the sunny South"*: His letter to the *Times-Picayune* was reprinted in an article called "The Octoroon Gone Home," *New York Times*, Feb. 9, 1860.

108 *The* Sunday Times *declared*: All the newspaper praise was collected and reprinted by Barnum in an ad in the *New-York Daily Tribune*, March 7, 1860.

109 *The truth, which the papers helpfully hid*: More details about William Henry Johnson can be found in Bogdan, *Freak Show*; and Adams, *E Pluribus Barnum*.

109 *"a certain museum proprietor in St. Louis"*: Cook, *Arts of Deception*.

CHAPTER 18. FIGHTING WORDS

110 *The famously gray skies*: Specific details of the Oxford event were collected from a variety of sources, including contemporary newspaper accounts and the transcripts of the speeches delivered at the event. Particularly useful were British Association for the Advancement of Science, *Proceedings of the Thirtieth Meeting, at Oxford* (London: John Murray, 1860); "The British Association for the Advancement of Science," *Lancet* 76 (1860); "The British Association," *Athenaeum*, July 7, 1860.

111 *In the twenty months after its publication*: Desmond and Moore, *Darwin*.

111 *"Let us ever apply ourselves"*: *Athenaeum*, July 7, 1860.

112 *"anxiety & consequent ill health"*: Darwin to Asa Gray, July 3, 1860.

112 *"Professor Owen wished to approach this subject"*: *Athenaeum*, July 7, 1860.

113 *Huxley "denied altogether that the difference"*: Ibid.

113 *But as several modern historians have since pointed out*: A very good analysis and debunking of the myths associated with the meeting can be found in Keith Thomson, "Huxley, Wilberforce, and the Oxford Museum," *American Scientist*, May–June 2000.

114 *"the irresistible tendency of organized beings"*: *Athenaeum*, July 7, 1860.

114 *"This gave Huxley the opportunity"*: Wollaston, *Life of Alfred Newton.*

115 *"I think the Bishop had the best of it"*: Balfour Stewart to James David Forbes, July 4, 1860.

115 *"uglyness & emptyness & unfairness"*: Hooker to Darwin, July 2, 1860.

115 *"I think I thoroughly beat him"*: Wilberforce to Sir Charles Anderson, July 3, 1860.

116 *"My dear Sir, let me present you"*: The letter is quoted in Owen, *Life*, vol. 2. The first letter that Du Chaillu sent to Owen was on Dec. 21, 1861, in which he informed Owen of his intent to place his specimens "at your service" (Rupke, *Richard Owen*).

CHAPTER 19. THE BOULEVARD OF BROKEN DREAMS

117 *Abraham Lincoln checked in to the Astor House*: Holzer, *Speech That Made Abraham Lincoln President.*

117 *photographic studio that was known as Broadway Valhalla*: Horan, *Mathew Brady.*

119 *"Ah," Lincoln said, "I see you want"*: Meredith, *Mr. Lincoln's Camera Man, Mathew B. Brady.*

119 *"No man ever before made such an impression"*: *New-York Tribune*, Feb. 28, 1860.

119 *"Brady and the Cooper Union made me president"*: Horan, *Mathew Brady.*

119 *"incertitude, and absolute darkness"*: Donald, *Lincoln.*

120 *"After viewing these monsters"*: "The Gorilla," *New York Post*, March 29, 1860.

120 *"in no way concerns the Academy"*: The Academy of Natural Sciences of Philadelphia Du Chaillu archive.

121 *"inexpedient to report the facts"*: Ibid.

121 *"A careful examination of its records shows"*: Ibid.

121 *He dropped the matter*: Although he stopped pursuing repayment from the academy at this time, Du Chaillu continued to complain to friends in letters throughout 1861 that the academy owed him money.

122 *Kneeland, who had extensive experience*: Vaucaire, *Gorilla Hunter*; and K. David Patterson, "Paul B. Du Chaillu and the Exploration of Gabon, 1855–1865," *International Journal of African Historical Studies* 7, no. 4 (1974).

122 *Wyman . . . had earlier written to his English counterpart*: Owen to Wyman, Nov. 1861.

CHAPTER 20. THE INNER CIRCLE

125 *2,803,921 human souls*: The population figure for 1861 comes from Pardon, *Popular Guide to London and Its Suburbs*.

125 *"London is the political, moral, physical"*: Ibid.

126 *Owen considered himself a self-made man*: Owen, *Life*, vol. 1. Owen's personality and positive attributes seemed to be more apparent to those outside his field. Some details of his physical bearing and lecture style have been drawn from the descriptions in McCarthy, *Portraits of the Sixties*.

127 *"Let them, if within their means"*: Owen, *Life*, vol. 1.

128 *The young gorilla hunter quickly became*: Ibid., vol. 2.

128 *The organization had added 233 fellows*: "Royal Geographical Society," *Times* (London), May 28, 1861.

128 *Captain William Sandbach, who ran a West Indies*: Sandbach had just been elected a fellow of the RGS when Du Chaillu arrived in early Feb. 1861. Sandbach had inherited a West India shipping company from his father called Sandbach, Tinné & Co. that traded in molasses, rum, and "prime Gold Coast Negroes," according to the Archives of London and the M25 Area. But the slave trade had been banned and stopped by the time William took over in 1851. The firm's primary trade was between England and Guyana.

129 *"My first impressions of your* Adventures": Murray to Du Chaillu, Feb. 19, 1861.

130 *a group dubbed the "four o'clock friends"*: Paston, *At John Murray's*.

130 *"the literary forum of the elite"*: Ibid.

130 *"Yours of this day has been received"*: Du Chaillu to Murray, Feb. 19, 1861.

CHAPTER 21. THE UNVEILING

132 *"The storms that had broken windows"*: *Times of London*, Feb. 25–31, 1861.

132 *The gas lamps flared*: In addition to details collected in newspaper reports and advertisements, I relied on several sources to re-create the scene of street life in the West End, including Thompson, *Visitor's Universal New Pocket Guide to London*; *Black's Guide to London and*

Its Environs (Edinburgh: Adam & Charles Black, 1863); Charles Dickens, "Arcadia," *Household Words* 20 (1853); Fyfe, *Images of the Street*; and Picard, *Victorian London*.

133 *a very particular breed of Londoner*: Some details on the crowd at Burlington House come from Dallas, *Series of Letters from London*.

133 *Inside Burlington House's west wing*: Building details from the diaries of George Mifflin Dallas, and also from Markham, *Fifty Years' Work of the Royal Geographical Society*.

134 *he stood about five feet three and weighed*: The physical descriptions of Du Chaillu are composed from numerous sources, including newspaper reports and Hills, *Author*, vol. 1.

134 *The crowd was thick with "savants"*: Francis Galton, *Proceedings of the Royal Geographical Society* 5 (1861).

135 *"the lowest schoolboy in the school"*: Ibid.

135 *"I trust he will offer me his protection"*: Dallas, *Series of Letters from London*.

135 *Animated gestures gave shape to his words*: Many newspaper reports about Du Chaillu in 1861 commented on his animated lecture style, as did Caroline Owen in her diary (Owen, *Life*, vol. 2).

136 *Murray's kids would beg the "Monkey Man"*: Paston, *At John Murray's*.

136 *"My children will never forget"*: Clodd, *Memories*.

136 *"almost the last man whom one at first sight"*: Alfred H. Guernsey, "Du Chaillu, Gorillas, and Cannibals," *Harper's Monthly*, April 1868.

137 *The audience was hooked*: The short article in the *Times* on Feb. 27 about the RGS meeting commented that "M. Du Chaillu's wanderings were told in a humorous style, and provoked a great deal of laughter." Other reports underscored the audience's unusual fascination with the lecture, including "The Gorilla Region of Africa," *Times* (London), March 5, 1861; "Royal Geographical Society," *Weekly Chronicle*, March 2, 1861.

137 *"the most strange and extraordinary animal"*: Galton, *Proceedings*.

138 *"In natural history, as we go on comparing"*: Ibid.

138 *There were dozens of ladies*: Dallas, *Series of Letters from London*; and Dallas, diaries.

138 *He had castrated them*: References to the castration can be found in Dawson, *Darwin, Literature, and Victorian Respectability*; Morris and Morris, *Men and Apes*; and Burton, *Two Trips to Gorilla Land*. In his

scandalous sixteen-volume translation of *The Book of the Thousand Nights and a Night* (Benares, India: Kamashastra Society, 1885–1888), Burton includes the following footnote, which vividly illustrates his reliable disregard of politesse: "[The] private parts of the monkey . . . are not of the girth sufficient to produce that friction which is essential to a woman's pleasure. I may here allude to the general disappointment in England and America caused by the exhibition of my friend Paul du Chaillu's Gorillas: he had modestly removed penis and testicles . . . and his squeamishness caused not a little grumbling and sense of grievance—especially amongst the curious sex" ("Supplemental Nights," 4:333n).

138 *"one of the oldest Fellows"*: *Times* (London), March 5, 1861.

139 *John Thadeus Delane, received a dinner invitation*: Dasent, *John Thadeus Delane, Editor of "The Times,"* vol. 2.

139 *"where, with amazing chemistry, Tom Towers"*: Trollope, *Warden*.

CHAPTER 22. THE GREAT WHITE HUNTER

141 *On a spring morning in Bloomsbury*: Interior descriptions of the Mudie library as it appeared in 1861 come from *Once a Week*, Dec. 21, 1861. More information about the business's background and operating methods is from Guinevere Griest, "Mudie's Circulating Library and the Victorian Novel," *Modern Philology* 69 (1972).

144 *Mudie, who'd had good luck with travel narratives*: Du Chaillu's popularity in Mudie's library comes from "Metropolitan Notes," *Journal of Education for Upper Canada* 12–14 (1861).

144 *It shot to the very top of Mudie's*: From advertisements in the *Athenaeum*.

144 *"We must go back to the voyages of Le Perouse"*: "The Discoveries of M. Du Chaillu," *Times* (London), May 20, 1861.

145 *"M. du Chaillu's narrative will not disappoint"*: The reviews originally appearing in the *Saturday Review*, the *Spectator*, and the *Critic* were reprinted in the *Spectator* 34 (1861).

147 *"I travelled—always on foot, and unaccompanied"*: Du Chaillu, *Explorations and Adventures*, viii.

147 *The book dripped with sensational descriptions*: For an analysis of the psychological interpretation of the book as a symbol of Europe's con-

ception of Africa, see Ben Grant, "'Interior Explorations': Paul Belloni du Chaillu's Dream Book," *Journal of European Studies* 38, no. 4 (2008).

148 *"I am sorry to be the dispeller"*: Du Chaillu, *Explorations and Adventures*, 347.

148 *"incontestable proof of the great ascendancy"*: Ibid.

149 *"I protest I felt almost like a murderer"*: Ibid., 60.

149 *"to find the very home of the beast"*: Ibid., 69.

149 *lecture hall that was "crowded to excess"*: Owen, *Life*, vol. 2.

149 *"M. Du Chaillu gave a very quaint"*: Ibid.

150 *"Shooting a lion"*: Quotation comes from an overview of the Royal Society lecture in the *Church of England Magazine*, May 11, 1861.

CHAPTER 23. INTO THE WHIRLWIND

151 *"from their size might fairly be denominated tusks"*: From a report in the *Caledonian Mercury* (Edinburgh, Scotland), July 9, 1861.

152 *But in 1843, London's* Punch *magazine*: "Cartoon No. 1," *Punch*, July 1843.

152 *"Am I a Man and a Brother?"*: "Monkeyana," *Punch*, May 18, 1861, 206. For more about the relationship between the evolution debate and cartoons, see Constance Areson Clark, "'You Are Here': Missing Links, Chains of Being, and the Language of Cartoons," *Isis*, Sept. 2009.

154 *"Monkeyana" was penned by Sir Philip Egerton*: Rupke, *Richard Owen*.

154 *"man of insignificant personal strength"*: "Discoveries of M. Du Chaillu."

155 *"But I left the note myself at your door"*: The Wilberforce story was repeated in several publications. The quotations are taken from "Our Foreign Bureau," *Harper's Magazine*, Vol. 23 (1861).

155 *"My name it is gorilla"*: Charles Handel Rand Marriott, "The Gorilla Quadrille," sheet music and lyrics obtained from the National Library of Australia.

156 *"The Lion of the Season"*: "The Lion of the Season," *Punch*, May 25, 1861, 213.

156 *Lent's wife, Julia Pastrana*: Several accounts of Pastrana's life were written shortly after her death. A modern perspective can be found in Janet Browne and Sharon Messenger, "Victorian Spectacle: Julia Pastrana, the Bearded and Hairy Female," *Endeavor* 27, no. 4 (Dec. 2003).

157 *"The figure," wrote the* Lancet: "A New Process of Embalming and Preserving the Human Body," *Lancet*, March 15, 1862.

157 *Dickens believed every word printed*: In *All the Year Round*, Dickens advised readers that every word of the magazine was "to be received as the statements and opinions of its Conductor." He called himself the "Conductor" of the magazine because he believed "editor" didn't convey the hands-on nature of his role. For more on Dickens's editing role, see Victor Sage, "Dickens and Professor Owen: Portrait of a Friendship," in *Le Portrait* (Paris: Presses de l'Université de Paris-Sorbonne, 1999).

157 *"If you knew how much interest it has awakened"*: Dickens to Owen, July 12, 1865.

158 *For Dickens, the difference seemed*: For more on the impact of the evolution debate on Dickens's writing, see Goldie Morgentaler, "Meditating on the Low: A Darwinian Reading of *Great Expectations*," *Studies in English Literature, 1500–1900* 38, no. 4 (1998). In reference to *Great Expectations*, Morgentaler cites the publication of Du Chaillu's book as one of several pieces of evidence that "traces of Darwinism should lie beneath the surface of Dickens's text."

158 *"The stupid weak savage"*: "An Ugly Likeness," *All the Year Round*, June 1, 1861, 237–40.

158 *"put a bad construction"*: "Next Door Neighbours to the Gorilla," *All the Year Round*, July 27, 1861, 423–27.

158 *"A gentleman of this disposition"*: Ibid.

159 The Coral Island—*an adventure about three English youths*: The enormously popular novel was, a century later, the inspiration for William Golding's *Lord of the Flies*, which subverts Ballantyne's romantic optimism.

159 *"hideous creatures one beholds"*: Ballantyne, *Gorilla Hunters*.

160 *he interpreted it as a veiled attack*: For more on Thackeray's response, see Cantor, *Science in the Nineteenth-Century Periodical*.

160 *"What do you think?"*: *Letters and Private Papers of William Makepeace Thackeray*, vol. 2.

160 *a satirical response printed in the* Cornhill Magazine: "Roundabout Papers," *Cornhill*, July 1861.

160 *London's police courts were clogged*: Hollingshead, *Ragged London in 1861*.

160 *She justified the beating*: Reade, *Savage Africa*.

161 *"Behold me here!" Byron announced*: "Savage Club," *Baily's Magazine of Sports and Pastimes*, vol. 3 (London: Baily Brothers, 1861).

161 *"Say, 'Am not I a savage and a brother?'"*: "The Savage Club," *Crosth-*

waite's Register of Facts and Occurrences Relating to Literature, the Sciences, and the Arts, July 1861.

162 *Every day of June on Regent Street*: Based on classified advertisements in the *Times* (London).

162 *The playbill advertised Paul J. Bedford*: The Adelphi Theatre Project, Calendar for 1860–1861.

162 *According to some calculations*: In David M. Wrobel, "Exceptionalism and Globalism: Travel Writers and the Nineteenth-Century American West," *Historian* 68, no. 3 (2006), sales for Du Chaillu's book were estimated at 300,000 copies.

162 *"excellent opportunity wasted by them"*: "Explorations and Travels," *National Quarterly Review* 3 (Sept. 1861): 393–97.

CHAPTER 24. THREE MOTIVES

164 *defining landmark had always been the pub*: Some of the atmosphere around the Elephant & Castle comes from "Some Things in London and Paris, 1836–1869," *Putnam's Magazine*, Jan.–June 1869; Dyos and Wolff, *Victorian City*; and Bacon, *Spurgeon*.

165 *"A monster place of worship"*: Cater, *Punch in the Pulpit*.

165 *"Some persons, you know, will not go"*: Mathews, *Hours with Men and Books*.

165 *"the social zone between the mechanic"*: Dowling, *London Town*, vol. 2.

166 *Spurgeon roamed the tabernacle's stage*: Bacon, *Spurgeon*.

166 *"I was told, and I believe, that in Agricultural Hall"*: William Cleaver Wilkinson, introduction to *Charles Haddon Spurgeon*, by Pike and Fernald.

166 *"as strong with pepper as can be borne"*: Ibid.

166 *"Let us commence the present service"*: Ibid.

167 *He devoured it, reading night and day*: C. H. Spurgeon's Autobiography.

167 *"December 19th was Sunday by my account"*: Du Chaillu, *Explorations and Adventures*.

167 *"Let me paint a set of slides on the gorilla"*: C. H. Spurgeon's Autobiography.

168 *To reach Walton Hall's front entrance*: The building, and the knockers, still exist today; the site is now part of the Waterton Park Hotel.

168 *Past the entrance to the dining room*: Some of the period details of

Walton Hall come from Hobson, *Charles Waterton, His Home, Habits, and Handiwork*; Blackburn, *Charles Waterton*; and Edginton, *Charles Waterton*.

169 *"a person recently discharged from prison"*: Blackburn, *Charles Waterton*.

169 *"a spider after a long winter"*: Ibid.

170 *"tapping the claret"*: Hobson, *Charles Waterton, His Home, Habits, and Handiwork*.

170 *John Edward Gray hadn't attended Paul's lectures*: J. E. Gray, "Zoological Notes on Perusing M. du Chaillu's *Adventures in Equatorial Africa*," *Annals of Natural History* 7 (1861).

CHAPTER 25. THE GORILLA WAR

172 *"Some time ago the arrival of a new African traveller"*: "The New Traveller's Tales," *Athenaeum*, May 18, 1861.

173 *"they have been preserved in or near the habitation"*: Ibid.

174 *The illustrations included in his book*: Vaucaire, *Gorilla Hunter*.

174 *"I hope that neither in my book"*: P. B. Du Chaillu, letter to the editor, *Times* (London), May 22, 1861.

175 *"This at least is certain"*: Paul Du Chaillu, "The New Traveller's Tales," *Athenaeum*, May 25, 1861.

176 *"If Mr. Du Chaillu had published his work"*: J. E. Gray, letter to the editor, *Times* (London), May 24, 1861.

176 *Gray's allies in the press proposed*: Despite Du Chaillu's later claims, Gray himself didn't propose the name change.

177 *"deliberately falsified material"*: Barth's article appeared in *Zeitschrift für allgemeine Erdkunde* 10 (1861): 430–67.

177 *When a new map of equatorial Africa*: The map was prepared by Dr. August Petermann and appeared in *Petermanns Geographische Mitteilungen* (Gotha: Justus Perthes, 1862).

177 *"As we were lazily sailing along"*: Du Chaillu, *Explorations and Adventures*.

177 *According to the story, someone asked him*: "M. Du Chaillu's Eagles," letters to the editor, *Times* (London), June 6, 1861.

178 *To celebrate the organization's anniversary*: "Royal Geographical Society," *Times* (London), May 28, 1861.

179 *"one of the boldest ventures which man"*: "Sir Roderick I. Murchison's Address," *Proceedings of the Royal Geographical Society* 5 (1861).

179 *"Strikingly attractive and wonderful"*: Ibid.

180 *"Whether one judges Monsieur Du Chaillu"*: *Times* (London), May 28, 1861.

180 *"I feel almost overwhelmed by the compliment"*: The *Times* (May 28, 1861) transcribed Du Chaillu's speech, complete with the audience's reaction, but rendered it in the past tense and the third person. I've altered the tenses and the pronouns to more naturally reflect Du Chaillu's words.

181 *"an uneducated collector of animal skins"*: Gray, "Zoological Notes."

181 *"He says they seem to have been wounded"*: J. E. Gray, "On the Habits of the Gorilla and Other Tailless Long-Armed Apes," *Proceedings of the Zoological Society of London*, May 28, 1861.

182 *"I then inquired of {the taxidermist}"*: "On the Death-Wound of the 'King of the Gorillas,'" letter read by Gray at the British Association for the Advancement of Science, Sept. 1861.

182 *Egerton insisted the evidence was entirely consistent*: "The Gorilla," *Athenaeum*, Sept. 21, 1861.

CHAPTER 26. THE SQUIRE'S GAMBIT

184 *"I immediately seized his forelegs"*: Waterton, *Wanderings in South America*.

184 *Those who knew him best*: See Hobson, *Charles Waterton*.

185 *"self-constituted censorious scoundrels"*: Gosse, *Squire of Walton Hall*.

185 *Waterton hadn't seen Paul's specimens*: Letters of Charles Waterton of Walton Hall.

186 *"It must have been on its hind legs"*: Ibid.

186 *"Our closet naturalists may gulp"*: Ibid.

187 *Mrs. Wombwell's Travelling Menagerie*: This traveling show was well known among nineteenth-century naturalists; Richard Owen, for one, occasionally received exotic specimens from it.

187 *"Having mounted the steps"*: "Watertonia," *Living Age* 56 (Jan.–March 1858).

188 *"She journeyed on"*: Ibid.

188 *Gorillas, he concluded, were made to swing*: Letters of Charles Waterton of Walton Hall.

189 *"scandalized beyond measure"*: Waterton, *Essays on Natural History.*

189 *"I allude to an occurrence"*: Hobson, *Charles Waterton.*

190 *"What a clever fellow Du Chaillu has been"*: Waterton to Mrs. W. Pitt Byrne, July 14, 1861.

CHAPTER 27. THE GORILLA IN THE PULPIT

191 *"It's nothing but the pictures"*: C. H. Spurgeon's *Autobiography.*

191 *"The Gorilla and the Land He Inhabits"*: In addition to the text of Spurgeon's lecture, several articles about the lecture were used to collect the descriptions and quotations in this chapter, including "Mr. Spurgeon on the Gorilla," *Liverpool Mercury,* Oct. 3, 1861; "Mr. Spurgeon on the Gorilla," *Times* (London), Oct. 3, 1861; "Mr. Spurgeon on the Gorilla," *Morning Post,* Oct. 4, 1861; "Annals of the Band of Hope Union," in *The Band of Hope Record, April 1861 to December 1862* (London: W. Tweedie, 1862); "Mr. Spurgeon on the Gorillas," *Literary Budget,* Nov. 1, 1861.

195 *"I can only say that if I travel again"*: Du Chaillu's comments were transcribed in the third person in the *Liverpool Mercury,* and I've changed them to the first person for clarity.

195 *wine, brandy, or ale was "absolutely necessary"*: Du Chaillu, *Explorations and Adventures,* 322.

196 *"to eat dirt and lick the shoes"*: "Mr. Spurgeon on the Gorillas," *Literary Budget.*

196 *"That the members of this church"*: C. H. Spurgeon's *Autobiography.*

197 *"This work of my Institution is of God"*: Ibid.

197 *For decades after this lecture*: For an analysis of Spurgeon's views on evolution, see Nigel Scotland, "Darwin and Doubt and the Response of the Victorian Churches," *Churchman* 100, no. 4 (1986).

197 *"Compromise there can be none"*: Drummond, *Spurgeon.*

CHAPTER 28. MRS. GRUNDY AND THE CANNIBAL CLUB

199 *Leicester Square was not a place*: In addition to period guidebooks of London, some details about Leicester Square were found in Lutz, *Pleasure Bound.*

199 *A small cadre of culture warriors*: For information on the Cannibal Club, I relied on letters and a disparate collection of books, including Brodie, *Devil Drives*; Sigel, *Governing Pleasures*; Henderson, *Swinburne*;

Swinburne, *Swinburne Letters*; Bercovici, *That Blackguard Burton!*; Farwell, *Burton*; and Kennedy, *Highly Civilized Man*.

200 *"Preserve us from our enemies"*: Only twenty copies of "The Cannibal Catechism" were printed in 1913 from a manuscript from Edward Gosse, a friend and biographer of Swinburne's.

201 *The name came from a play*: Morton's play was *Speed the Plough* (1798).

201 *"absolutely unfit for the Christian population"*: Brodie, *Devil Drives*.

201 *"Mrs. Grundy is already beginning to roar"*: Wright, *Life of Sir Richard Burton*.

202 *The forty-year-old Burton was at a crossroads*: I relied on numerous biographies, listed above, for the biographical sketch of Richard Burton; among the most helpful were those by Lovell, Rice, Brodie, and Farwell.

205 *who'd already mingled socially with Milnes*: The two men crossed paths at meetings of the Royal Society in early 1861, according to the organization's proceedings.

205 *"the governmental crumb"*: Farwell, *Burton*.

206 *"a most able paper which wanted nothing"*: Accounts of the feud are in Kennedy, *Highly Civilized Man*; and Baker, *History of Geography*.

207 *"gentlemen and players"*: Kennedy, *Highly Civilized Man*.

207 *"a sin of omission"*: Richard Burton, "Ethnological Notes on M. du Chaillu's *Explorations and Adventures in Equatorial Africa*," read before the Ethnological Society of London, 1861.

208 *The voice belonged to Thomas Malone*: Some information on Malone's history comes from "On Engraving by Light and Electricity," *Journal of the Franklin Institute* 64 (1857); "Lectures on Photography," *Photographic Journal* 3 (1857). The Database of 19th Century Photographers and Allied Trades in London, 1841 to 1901, accessed at www.photolondon. org, also provided some biographical details.

208 *He had not read* Explorations and Adventures: T. A. Malone, letter to the editor, *Times* (London), July 5, 1861.

209 *"Did you see everything you describe"*: Malone's questions were alluded to in the articles about the event, including his own letter to the *Times* referenced above.

209 *"Of course all my remarks were unpalatable"*: Ibid.

209 *"He then rose"*: Richard F. Burton, letter to the editor, *Times* (London), July 8, 1861.

209 *"Soon after the chairman had left"*: James Hunt, letter to the editor, *Times* (London), July 8, 1861.

209 *Several witnesses heard Paul shout*: *Spectator*, July 6, 1861; "Ethnological Society," *Lancet*, July 6, 1861.

209 *"I was preparing to go"*: Malone, letter to the editor.

210 *Feeling stifled by the "respectability"*: Wright, *Life of Sir Richard Burton*.

210 *"My wonder is that M. Du Chaillu"*: Burton, letter to the editor.

210 *"We fear M. Du Chaillu has been too long"*: "Mr. Du Chaillu and His Detractors," *Examiner*, July 6, 1861.

CHAPTER 29. "EVIDENCE OF A SPURIOUS ORIGIN"

212 *Alexander Wilson, a Scotsman whose* American Ornithology: Edginton, *Charles Waterton*; and Blackburn, *Charles Waterton*.

213 *"It is then that you may see the cruel spirit"*: Rhodes, *John James Audubon*.

213 *They once correctly called Audubon out*: Edginton, *Charles Waterton*.

214 *"Audubon has been rudely assailed"*: Arthur, *Audubon*.

214 *He went so far as to surrender his claims*: Rhodes, *John James Audubon*.

214 *"Without leaving behind him in America"*: Charles Waterton, "Remarks on Audubon's Biography of Birds," *Magazine of Natural History and Journal of Zoology, Botany, Mineralogy, Geology, and Meteorology*, vol. 6 (London: Longman, Rees, Orme, Brown, and Green, 1833).

215 *"many of those who have afforded their patronage"*: Souder, *Under a Wild Sky*.

215 *"Audubon is immaculate when compared"*: Waterton to Ord, June 27, 1861.

215 *"I have warmed his hide"*: Ibid.

216 *"If it be a fact that he is a mongrel"*: Ord to Waterton, Oct. 1861.

CHAPTER 30. SHADOWS OF THE PAST

217 *he'd hint that there were Huguenots*: Several biographical entries about Du Chaillu in encyclopedias list him as being born of French Huguenot parents in New Orleans.

217 *People naturally assumed that "Belloni"*: Vaucaire, *Gorilla Hunter*.

218 *Edward Clodd, an eminent banker*: Clodd, *Memories*.

218 *The island of Réunion*: Hombert and Perrois, *Cœur d'Afrique*.

219 *children born of a white man*: Bucher, "Canonization by Repetition."

219 *Charles-Alexis Du Chaillu was earning*: Du Chaillu's father's activities on the island come from "Les 'Francs-Créoles,'" *Journal de l'Île de la Réunion*, Jan. 1, 2005; and Hombert and Perrois, *Cœur d'Afrique*.

221 *a ship registered to Charles-Alexis was detained*: Many of the records concerning Charles-Alexis, including his relationship with Bréon, are from Annie Merlet, "Paul Belloni Du Chaillu; ou, L'invention d'un destin," which is included in Hombert and Perrois, *Cœur d'Afrique*.

222 *"He hated the country in which such things"*: Evertson Smith, "Reminiscences."

222 *Charles-Alexis was in charge*: Bucher, "Canonization by Repetition." Du Chaillu's father continued to show up in administrative documents until the end of 1855, and he is believed to have died then (Burton, *Two Trips to Gorilla Land*). However, Du Chaillu once wrote that his father died in 1851 and that he, Paul, traveled from France to Gabon at that time to take care of his father's affairs—a statement that is contradicted by the Gabonese records and other statements by Du Chaillu concerning his father.

223 *According to Putnam County, New York, records*: Ibid.

224 *"M. du Chaillu is a bald, bronzed"*: "Central Africa," *New York World*, May 9, 1867.

224 *"By the way, it seems to be another disputed"*: "Affairs in London," *New York Times*, July 19, 1861.

224 *"written by an old enemy of his and sent to me"*: Clodd, *Memories*. Some have speculated that R. B. N. Walker was the author of the unpublished manuscript (see Nora McMillan, "Robert Bruce Napoleon Walker," *Archives of Natural History* 23, no. 1 [1996]).

225 *"Men hear gladly of the power of blood"*: Emerson, *English Traits*.

CHAPTER 31. BLACK AND WHITE

226 *"I met every where in my travels"*: Du Chaillu, *Explorations and Adventures*, 21.

227 *significantly less offensive than others*: Livingstone also was generally less dismissive of native Africans than other travelers of the mid-nineteenth century.

227 *assumed that blacks were "less evolved"*: Often scientists supported the

idea by citing general differences in cranial capacity as evidence—differences that were hotly debated in the following centuries.

227 *"They roar with laughter"*: Du Chaillu, *Explorations and Adventures*, 284.

228 *"Now he belongs to the ages"*: Some have argued that Stanton actually said, "Now he belongs to the angels" (Gopnik, *Angels and Ages*).

228 *"Du Chaillu was a fool to wander"*: McClellan, *McClellan's Own Story*.

228 *"Since the Southerners have adopted the habit"*: *White Cloud Kansas Chief*, Aug. 8, 1861.

229 *"We do believe that if the African Gorilla"*: Van Tassel, *"Behind Bayonets."*

229 *"these human gorillas to murder"*: Greeley, *American Conflict*.

229 *"hopelessly degraded intellectual organization"*: MacMahon, *Cause and Contrast*.

230 *"In this there seems to be a palpable contradiction"*: Wheat, *Progress and Intelligence of Americans*.

232 *"a Nietzschean confrontation"*: Jeal, *Stanley*.

232 *"independence of mind"*: *Autobiography of Sir Henry Morton Stanley*.

232 *"neither fish nor fowl"*: Lady Isabel Burton, *The Life of Captain Sir Richard F. Burton* (London: Chapman & Hall, 1893).

232 *"For the half-castes I have great contempt"*: Stanley, *How I Found Livingstone*.

233 *"degraded the superior race"*: *Cincinnati Daily Enquirer*, Feb. 14, 1869.

CHAPTER 32. THE IMPOSTORS

234 *"the truth would right itself"*: "Royal Geographical Society," *Times* (London), May 28, 1861.

234 *Paul had written to the Presbyterian mission*: "Du Chaillu Vindicated," *Times* (London), June 3, 1862.

235 *"Letters came to town from the Gaboon"*: *Athenaeum*, Sept. 14, 1861.

235 *But the* Athenaeum *item was puzzling*: Du Chaillu to John Murray, Sept. 15, 1861.

235 *"Everything will be right"*: Ibid.

236 *"I, in common with most persons"*: The *Morning Advertiser* letter was reprinted in the *Athenaeum*, Sept. 21, 1861.

236 *Unless the letter wasn't really from*: Du Chaillu to Murray, Sept. 17, 1861.

236 *He'd moved from Sussex to Gabon*: Biographical information from McMillan, "Robert Bruce Napoleon Walker."

237 *"My enemies will not let me rest"*: Du Chaillu to Murray, Sept. 17, 1861.

237 *"Having known M. Du Chaillu for some years"*: *Athenaeum*, Sept. 21, 1861.

238 *"for the sake of greater courtesy"*: Ibid.

238 The Times, *for its part, declined*: Vaucaire, *Gorilla Hunter*.

239 *"An enterprising naturalist with whom"*: Walker to P. L. Simmonds, Nov. 4, 1858.

239 *progenitor of the black race*: John P. Daly, book review of *Noah's Curse: The Biblical Justification of American Slavery* by Stephen R. Haynes, *Journal of Southern History* 69, no. 4 (Nov. 2003).

239 *"Mr. Paul Duchaillu, the West African"*: Walker to Simmonds, May 3, 1859.

240 *"I cannot express to you the sorrow"*: Du Chaillu to Murray, Sept. 18, 1861.

240 *Within months, R. B. N. Walker would board*: "Gorillas in Liverpool," *Times* (London), June 3, 1862.

241 *"I am going to pen a few lines"*: McMillan, "Robert Bruce Napoleon Walker."

CHAPTER 33. SHORTCUTS TO GLORY

242 *"Belloni, it appears, is the traveler's"*: "M. Du Chaillu, His Book an Alleged Imposter," *Glasgow Examiner*, Feb. 13, 1862.

242 *"I simply told what I saw"*: "M. Du Chaillu's Lecture on the Gorilla," *Glasgow Herald*, Oct. 12, 1861.

243 Charlotte and Myra, *was dismissed*: The reviews of Reade's books come from Allibone and Kirk, *Critical Dictionary of English Literature and British and American Authors*.

243 *"discovered a short cut to glory"*: Reade, *African Sketch-Book*.

244 *"In my humble character as a mere collector"*: Reade, *Savage Africa*.

244 *"all of whom were tolerably"*: Ibid.

244 *Walker took one look at Reade*: From diaries of the Reverend William Walker, Wisconsin Historical Society.

244 *"As you have not been in a hot country"*: Reade, *Savage Africa*.

244 *"He is very social, & likes"*: Letter from Walker, Feb. 20, 1862.

246 *"I must shoot a gorilla"*: Reade to Walker, April 20, 1862.

246 *"His answer was precise"*: W. Winwood Reade, "The Gorilla as I Found Him," *Every Saturday*, Aug. 31, 1867, 270.

246 *"He had only shot little birds"*: Ibid.

247 *"Having spent five active months in the Gorilla Country"*: W. Winwood Reade, "News from the Gorilla Country," *Athenaeum*, Nov. 22, 1862.

248 *"Thus in an obscure African village"*: Ibid.

CHAPTER 34. THE WAGER

249 *"What are five months to traverse"*: Du Chaillu, letter to the editor, *Times* (London), Dec. 1, 1862.

250 *"The old African hunters will not take"*: Ibid.

250 *"I will prove how I got my specimens"*: Ibid.

251 *"a sum much below that at which"*: "Du Chaillu's Collection," *Times* (London), June 13, 1863.

CHAPTER 35. THE REINVENTION

256 *"addressed to a person who"*: The handbook was first printed as an insert in the *Journal of the Royal Geographical Society*, vol. 24 (London: John Murray, 1854).

256 *see entries like "bones as fuel"*: Galton, *Art of Travel*.

256 *"With such a book in his hand"*: Marcy, *Prairie Traveler*.

256 *Richard Burton, for one, rarely traveled*: Rice, *Captain Sir Richard Francis Burton*; and Kennedy, *Highly Civilized Man*.

257 *R. B. N. Walker reported to the* Times: R. B. N. Walker, letter to the editor, *Times* (London), Dec. 5, 1862.

257 *"I will briefly state that, after a residence"*: Richard Burton, letter to the editor, *Times* (London), Dec. 23, 1862.

258 *revising* Hints to Travellers *for a new 1864 edition*: Markham, *Fifty Years' Work of the Royal Geographical Society*.

258 *Back, who tutored him*: Their friendship is evident in numerous letters between the two men held by the RGS, and Du Chaillu thanked him in *Journey to Ashango-Land*.

258 *"We were two days about the ascent"*: Du Chaillu, *Explorations and Adventures*, 442.

259 *As he studied with Back, Paul also took*: Information about Du

Chaillu's various apprenticeships from letters to Back; Vaucaire, *Gorilla Hunter*; and Du Chaillu, *Journey to Ashango-Land*.

260 *On August 6, 1863, a hundred-ton schooner*: Du Chaillu, *Journey to Ashango-Land*; and Du Chaillu, *Country of the Dwarfs*.

260 *Some of the boxes were packed tight*: Descriptions of equipment from Du Chaillu, *Country of the Dwarfs*, and letters from Du Chaillu to John Murray sent from Gabon in 1863 and 1864.

261 *"Be assured that I will apply the amount"*: Du Chaillu to Murray, June 22, 1863.

261 *The resulting shopping spree*: Du Chaillu to Murray, Jan. 2, 1864; and Du Chaillu, *Journey to Ashango-Land*.

262 *"to fix with scientific accuracy"*: Du Chaillu, *Journey to Ashango-Land*, 1.

CHAPTER 36. DAMAGED GOODS

263 *The* Mentor *lingered a teasing distance*: The account of the troubled entry onto land was reconstructed from Du Chaillu, *Journey to Ashango-Land*, and letters from Du Chaillu to Murray and Back written in 1863 and 1864.

264 *"with great exertions kept me"*: Du Chaillu, *Journey to Ashango-Land*, 11.

265 *To help defray the costs*: Du Chaillu to William Walker, April 15, 1864.

266 *"I like to practice while I am near"*: Du Chaillu to Murray, Nov. 15, 1863.

266 *"I felt and still feel the warmest friendship"*: Du Chaillu, *Journey to Ashango-Land*, 18.

266 *"The old fellow was so delighted"*: Du Chaillu to Murray, Jan. 2, 1864.

267 *hearing Tom's "frantic cries"*: "Amidst the Ruins," *Hardwicke's Science-Gossip*, vol. 3 (1868).

268 *"there was a strong party"*: Du Chaillu, *Journey to Ashango-Land*, 66.

269 *"Once or twice they seemed on the point"*: Ibid., 50.

269 *"The huge beast stared at me"*: Ibid.

CHAPTER 37. THE BOLDEST VENTURE

272 *"If I looked at him he would make a feint"*: Du Chaillu, *Journey to Ashango-Land*, 68.

272 *"Her death was like that of a human"*: Ibid., 55.

272 *"I had sent him consigned to Messrs. Baring"*: Du Chaillu wrote of the consignment in letters to Murray, to whom he sent the money to buy the insurance. Among the dangers that Du Chaillu feared the ship might face was war. In Africa, Du Chaillu had been told by crew members of an English ship that both France and England had declared war against America, turning the U.S. Civil War into an international battle. "I hope the Yankees will not get hold of the Mentor," Du Chaillu wrote to Murray. "It will be a case."

273 *requesting a copy of the book* Cosmos: Du Chaillu to Murray, Jan. 14, 1864.

274 *"chain of connection, by which all natural forces"*: Humboldt, *Cosmos*.

274 *"troops of marauders, who roam over the steppes"*: Humboldt, *Personal Narrative of Travels to the Equinoctial Regions of America*.

274 *Ralph Waldo Emerson and Henry David Thoreau*: In his intellectual biographies of both Emerson (*Emerson: The Mind on Fire* [Berkeley: University of California Press, 1995]) and Thoreau (*Henry Thoreau: A Life of the Mind* [Berkeley: University of California Press, 1986]), Robert D. Richardson noted their readings of Humboldt.

275 *"It was not my object on the present journey"*: Du Chaillu, *Journey to Ashango-Land*, ix.

275 *"I am perfectly tired of this Gorilla business"*: Du Chaillu to Dr. Henry Bence Jones, Aug. 20, 1864.

276 *"In a few days I am off for the interior"*: Du Chaillu to Back, Aug. 20, 1864.

277 *"It is so good, so superior"*: Letter is quoted in Owen, *Life*, vol. 2.

277 *"In his last letter to me"*: "Sir Roderick I. Murchison's Address," *Proceedings of the Royal Geographical Society* 9 (1865).

CHAPTER 38. THE ARMIES OF THE PLAGUE

278 *About fifty men accompanied Paul*: Details of the expedition from Du Chaillu, *Journey to Ashango-Land*, and his account of the expedition delivered to the RGS, "Second Journey into Equatorial Africa," *Proceedings of the Royal Geographical Society* 10 (1866).

280 *"Macondai cursed the* okenda i nialai": Du Chaillu, *Journey to Ashango-Land*, 108.

281 *"They began boldly to accuse me"*: Ibid., 122.

282 *"Not a day passed"*: Ibid., 129.

282 *"Those who were now well enough"*: Ibid., 132.

284 *"I found it impossible to keep them all"*: Ibid., 146.

285 *"On the 1st and 3rd of April I over-exerted myself"*: Ibid., 163.

286 *"I told you when I came"*: Ibid., 179.

287 *"My misfortunes will never terminate!"*: Du Chaillu's journal transcriptions were included in *Journey to Ashango-Land*, 191.

CHAPTER 39. RUNNING FOR THEIR LIVES

289 *"We must go forward"*: Du Chaillu, *Journey to Ashango-Land*, 246.

289 *known as the Babongo people*: Today, many of the Babongo still live in the forested areas where Du Chaillu found them, and they are still renowned for their tracking skills. At the Max Planck gorilla research station in the Fernan-Vaz region, Josephine Head employs Babongo tribesmen, hired from inland communities, to work as professional trackers.

290 *"My {Ashango} guides were kind"*: Ibid., 324.

290 *"The musical box was brought out"*: Ibid., 305.

291 *"It is a tremendous task"*: Journal entry transcribed in ibid., 301.

293 *"I felt that it was time to make a stand"*: Ibid., 356.

296 *"You have no idea of the trials"*: Du Chaillu to Murray, Sept. 29, 1865.

CHAPTER 40. THE JURY OF HIS PEERS

297 *Paul, the featured speaker*: "Second Journey into Equatorial Africa," *Proceedings of the Royal Geographical Society* 10 (1866).

299 *Some of the people in the crowd*: "Discussion on M. Du Chaillu's Paper," *Proceedings of the Royal Geographical Society* 10 (1866).

299 *"deposited it in the hands"*: Ibid.

299 *An article in the* Transactions: Ibid., and "Prof. Allman on *Potamogale velox*," *Transactions of the Zoological Society of London* (1867).

300 *French exploratory boats had been pushing up*: Kingsley, *Travels in West Africa*; Patterson, "Paul B. Du Chaillu."

300 *The German cartographers who earlier*: Vaucaire, *Gorilla Hunter*.

300 *"astonished at the multitude and accuracy"*: "Discussion on M. Du Chaillu's Paper."

300 *"After these opportunities of further observation"*: "Second Journey."

301 *"I will only add that if M. Du Chaillu's"*: W. Winwood Reade, letter to the editor, *Times* (London), Jan. 23, 1866.

CHAPTER 41. ACCIDENTAL VICTORIES

302 *In early 1866, Owen published a paper*: Rupke, *Richard Owen*. The paper was "On the Anatomy of Vertebrates."

302 *Owen suggested that Darwin*: Ibid.

303 *"What an unmerciful basting you give"*: Huxley to Darwin, Nov. 11, 1866.

303 *Owen had published a monograph*: Owen, *Memoir on the Gorilla*.

304 *"men differ more widely from one another"*: T. H. Huxley, *Evidence as to Man's Place in Nature* (London: D. Appleton, 1863).

304 *"I have observed individuals of the Negro race"*: "Professor Owen on the External Characters and Affinities of the Gorilla," *Transactions of the Zoological Society of London* 5 (1866).

305 *"the respect and gratitude of every genuine"*: Du Chaillu, *Journey to Ashango-Land*, 439.

CHAPTER 42. THE EXPLORER

306 *Paul took full advantage of the offer*: An account of Paul's days in Twickenham appears in Burnand, *Records and Reminiscences*, vol. 2.

306 *He often worked alongside Henry Walter Bates*: Du Chaillu to Murray, Sept. 26, 1866.

307 *"a most amusing man, and very excitable"*: Burnand, *Records and Reminiscences*, vol. 2.

308 *"Nothing is more characteristic in his book"*: "Ashango Land," *Chambers's Journal of Popular Literature, Science, and Arts*, June 8, 1867.

308 *"You have all heard how"*: "Discussion on M. Du Chaillu's Paper."

CHAPTER 43. NO-MAN'S-LAND

312 *"They think it will have immense sale here"*: Du Chaillu to Murray, May 21, 1867.

312 *five popular adventure books*: Lysle E. Meyer, in his book *Farther Frontier*, stated that based on the sales and reprintings of these volumes, it was reasonable to conclude that Du Chaillu "had more to do with

Americans' conception of Africa for over a half century than any other author."

313 *"I used to be ashamed of hating him"*: Darwin to Joseph Dalton Hooker, Aug. 4, 1872.

313 *"The reader must bear in mind"*: Owen, *Life*, vol. 2.

314 *"It may be truth"*: Huxley, *Evidence*.

314 *"One of them says, after returning"*: Barrie, *Window in Thrums*.

314 *a close friend of Barrie's*: Du Chaillu even played, occasionally, on a cricket team formed by Barrie and called the Allahakbarries—a play on the Arabic phrase *allah akbar*.

314 *"I have been unable to find the slightest trace"*: Vaucaire, *Gorilla Hunter*.

315 *the Philafrican Liberators' League*: "Slaves' Unknown Friends," *New York Times*, Sept. 20, 1896.

316 *"She was so motherly to me"*: Ibid.

316 *When Judge Daly died*: "Will of Ex-Justice Daly," *New York Times*, Oct. 24, 1899.

317 *"marked weight with the Scandinavians"*: "Paul Du Chaillu in Politics," *Saint Paul Globe*, Aug. 11, 1896.

317 *Paul lobbied to land the position*: Du Chaillu to Charles Daly, May 19, 1897.

317 *"They are a grand, noble race"*: "P. B. Du Chaillu," *New York Times*, Dec. 30, 1873.

318 *"Mr. du Chaillu's books and explorations"*: "Not Weary of Travel," *Washington Post*, March 24, 1901.

319 *"I am restless, curious to see things"*: *Marion Daily Star*, May 20, 1901.

319 *employing two teachers for lessons*: Vaucaire, *Gorilla Hunter*.

319 *"I hope good health will be granted"*: Ibid.

319 *"I can't see," he said*: "Du Chaillu's Last Hours," *New York Times*, May 23, 1903.

319 *"I am dying"*: Ibid.

320 *"Have the remains embalmed"*: Ibid.

320 *He had given away almost everything*: "Du Chaillu Died Poor," *New York Times*, July 5, 1903.

320 *His body arrived in New York*: "Du Chaillu's Body Here," *New York Times*, June 19, 1903.

320 *"Now that the earthly ending has come"*: "Funeral of Paul Du Chaillu," *New York Times*, June 24, 1903.

321 *"As an explorer, as lecturer, as author"*: Ibid.

EPILOGUE

324 *Paul Du Chaillu had accompanied Doyle*: Redmond, *Welcome to America, Mr. Sherlock Holmes.*

325 *When London died in 1916*: *Jack London Newsletter* 21 (1988); and Alex Kershaw, *Jack London: A Life* (New York: Thomas Dunne Books, 1999).

325 *"I made up my mind right then"*: Interview with Merian Cooper by Rudy Behlmer, L. Tom Perry Special Collections, Harold B. Lee Library, Brigham Young University.

327 *"This was unfortunate, for he was basically"*: Schaller, *Year of the Gorilla.*

328 *"allow us to perceive through him a historical"*: Hombert and Perrois, *Cœur d'Afrique.*

329 *One of his offspring, a boy named André*: McMillan, "Robert Bruce Napoleon Walker."

329 *Raponda Walker perused the writings*: Walker, *Notes d'histoire du Gabon.*

Selected Bibliography

Ackroyd, Peter. *Dickens*. New York: HarperCollins, 1990.

Adams, Bluford. *E Pluribus Barnum: The Great Showman and the Making of U.S. Popular Culture*. Minneapolis: University of Minnesota Press, 1997.

Ade Ajayi, J. F., ed. *UNESCO General History of Africa*. Vol. 6, *Africa in the Nineteenth Century Until the 1880s*. Berkeley: University of California Press, 1989.

Aicardi de Saint-Paul, Marc. *Gabon: The Development of a Nation*. London: Routledge, 1989.

Allibone, Samuel, and John Foster Kirk. *A Critical Dictionary of English Literature and British and American Authors*. Philadelphia: J. B. Lippincott, 1899.

Arthur, Stanley Clisby. *Audubon: An Intimate Life of the American Woodsman*. Gretna, La.: Pelican, 1999.

Asbury, Herbert. *Gangs of New York: An Informal History of the Underworld*. New York: Vintage, 2008.

Bacon, Ernest. *Spurgeon: Heir of the Puritans*. Arlington Heights, Ill.: Christian Liberty Press, 1996.

Baker, J. N. L. *The History of Geography*. Oxford: B. Blackwell, 1963.

Ballantyne, R. M. *The Gorilla Hunters*. London: T. Nelson and Sons, 1861.

Barnum, P. T. *Struggles and Triumphs; or, Forty Years' Recollections of P. T. Barnum*. New York: Penguin Books, 1981.

Barrie, J. M. *A Window in Thrums*. New York: C. Scribner's Sons, 1896.

Barth, Heinrich. *Travels and Discoveries in North and Central Africa, Being a Journal of an Expedition Under the Auspices of H.B.M.'s Government*. New York: Harper & Brothers, 1857.

Beeckman, Daniel. *A Voyage to and from the Island of Borneo*. London: Dawsons of Pall Mall, 1973.

Bercovici, Alfred. *That Blackguard Burton!* Indianapolis: Bobbs-Merrill, 1962.

Blackburn, Julia. *Charles Waterton: Traveller and Conservationist*. London: Bodley Head, 1989.

Blaikie, William Garden. *The Life of David Livingstone*. London: John Murray, 1903.

Blake, William J. *The History of Putnam County, N.Y.* New York: Baker & Scribner, 1849.

Bobo, William M. *Glimpses of New York City*. Charleston, S.C.: J. J. McCarter, 1852.

Bogdan, Robert. *Freak Show: Presenting Human Oddities for Amusement and Profit*. Chicago: University of Chicago Press, 1990.

Bowler, Peter J. *Evolution: The History of an Idea*. Rev. ed. Berkeley: University of California Press, 1989.

———. *Monkey Trials and Gorilla Sermons: Evolution and Christianity from Darwin to Intelligent Design*. Cambridge, Mass.: Harvard University Press, 2007.

Brodie, Fawn M. *The Devil Drives: A Life of Sir Richard Burton*. New York: W. W. Norton, 1967.

Browne, Janet. *Charles Darwin: A Biography.* New York: Knopf, 1995.

———. *Darwin's "Origin of Species": A Biography.* Vancouver: Douglas & McIntyre, 2006.

Burnand, Sir Francis Cowley. *Records and Reminiscences, Personal and General.* Vol. 2. London: Methuen, 1904.

Burton, Richard F. *The Lake Regions of Central Africa.* London: Longman, Green, Longman, and Roberts, 1860.

———. *Two Trips to Gorilla Land and the Cataracts of the Congo.* London: S. Low, Marston, Low, and Searle, 1876.

Caldecott, Julian, and Lera Miles, eds. *World Atlas of Great Apes and Their Conservation.* Berkeley: University of California Press, 2005.

Cantor, Geoffrey. *Science in the Nineteenth-Century Periodical.* Cambridge, U.K.: Cambridge University Press, 2004.

Cater, Philip. *Punch in the Pulpit.* London: William Freeman, 1863.

Clarke, Michael Tavel. *These Days of Large Things: The Culture of Size in America.* Ann Arbor: University of Michigan Press, 2007.

Clodd, Edward. *Memories.* New York: Putnam, 1916.

Columbus, Christopher. *The Journal of Christopher Columbus.* Translated by Cecil Jane. London: Blond and the Orion Press, 1960.

Conan Doyle, Arthur. *The Lost World.* Oxford: Oxford University Press, 1995.

Cook, James. *The Arts of Deception: Playing with Fraud in the Age of Barnum.* Cambridge, Mass.: Harvard University Press, 2001.

Cooley, William Desborough. *Inner Africa Laid Open.* London: Longman, Brown, Green, and Longmans, 1852.

Cosans, Christopher E. *Owen's Ape and Darwin's Bulldog: Beyond Darwinism and Creationism.* Bloomington: Indiana University Press, 2009.

Crone, G. R. *Modern Geographers: An Outline of Progress in Geography Since A.D. 1800.* London: Royal Geographical Society, 1951.

Dallas, George Mifflin. *A Series of Letters from London.* Philadelphia: J. B. Lippincott, 1869.

Darwin, Charles. *The Descent of Man and Selection in Relation to Sex.* London: John Murray, 1871.

———. *On the Origin of Species.* London: John Murray, 1859.

———. *The Voyage of the* Beagle. London: P. F. Collier, 1909.

Dasent, Arthur Irwin. *John Thadeus Delane, Editor of "The Times": His Life and Correspondence.* Vol. 2. London: John Murray, 1908.

Dawson, Gowan. *Darwin, Literature, and Victorian Respectability.* Cambridge, U.K.: Cambridge University Press, 2007.

Desmond, Adrian, and James Moore. *Darwin.* London: Michael Joseph–Penguin Group, 1991.

Dickens, Charles. *American Notes for General Circulation.* New York: Penguin Classics, 2001.

———. *Great Expectations.* New York: Penguin Books, 1989.

Donald, David Herbert. *Lincoln.* New York: Simon & Schuster, 1996.

Dowling, Richard. *London Town: Sketches of London Life and Character, by Marcus Fall.* Vol. 2. London: Tinsley Brothers, 1880.

Drummond, Lewis A. *Spurgeon: Prince of Preachers.* Grand Rapids: Kregel, 1992.

DuBose, Hampden. *Memoirs of Rev. John Leighton Wilson.* Richmond: Presbyterian Committee of Publication, 1895.

Du Chaillu, Paul B. *The Country of the Dwarfs.* New York: Harper & Brothers, 1872.

———. *Explorations and Adventures in Equatorial Africa.* London: John Murray, 1861.

―――. *A Journey to Ashango-Land: And Further Penetration into Equatorial Africa*. London: John Murray, 1867.

―――. *King Mombo*. New York: Charles Scribner's Sons, 1902.

―――. *The Land of the Midnight Sun: Summer and Winter Journeys Through Sweden, Norway, Lapland, and Northern Finland*. New York: Harper & Brothers, 1882.

―――. *Stories of the Gorilla Country*. New York: Harper & Brothers, 1870.

―――. *The Viking Age: The Early History, Manners, and Customs of the Ancestors of the English-Speaking Nations*. New York: Charles Scribner's Sons, 1889.

Dyos, Harold James, and Michael Wolff. *The Victorian City: Images and Realities*. 2 vols. London: Routledge & Kegan Paul, 1973.

Edginton, Brian W. *Charles Waterton: A Biography*. Cambridge, U.K.: Lutterworth Press, 1996.

Emerson, Ralph Waldo. *English Traits*. London: G. Routledge, 1857.

Erb, Cynthia Marie. *Tracking King Kong: A Hollywood Icon in World Culture*. Detroit: Wayne State University Press, 2009.

Fabian, Johannes. *Out of Our Minds: Reason and Madness in the Exploration of Central Africa*. Berkeley: University of California Press, 2000.

Fage, J. D., and Roland Oliver, eds. *The Cambridge History of Africa*. Cambridge, U.K.: Cambridge University Press, 1975–1986.

Farwell, Byron. *Burton: A Biography of Sir Richard Francis Burton*. London: Holt, Rinehart and Winston, 1963.

Forster, John. *The Life of Charles Dickens*. London: Palmer, 1928.

Fossey, Dian. *Gorillas in the Mist*. New York: Mariner Books, 2002.

Franey, Laura E. *Victorian Travel Writing and Imperial Violence: British Writing on Africa, 1855–1902*. New York: Palgrave Macmillan, 2003.

Fyfe, Nicholas. *Images of the Street*. London: Routledge, 1998.

Galton, Francis. *The Art of Travel; or, Shifts and Contrivances Available in Wild Countries.* London: John Murray, 1855.

Geikie, Archibald. *Life of Sir Roderick Murchison.* London: John Murray, 1875.

Gladstone, W. E. *The Gladstone Diaries.* Oxford: Clarendon Press, 1968.

Godsall, Jon R. *The Tangled Web: A Life of Sir Richard Burton.* London: Matador, 2008.

Gopnik, Adam. *Angels and Ages: A Short Book About Darwin, Lincoln, and Modern Life.* New York: Knopf, 2009.

Gosse, Philip. *The Squire of Walton Hall: The Life of Charles Waterton.* London: Cassell, 1940.

Grant, Ben. *Postcolonialism, Psychoanalysis, and Burton: Power Play of Empire.* Hoboken, N.J.: Taylor & Francis, 2008.

Greeley, Horace. *The American Conflict: A History of the Great Rebellion in the United States, 1860–1865.* Hartford, Conn.: O. D. Case, 1866.

Griest, Guinevere. *Mudie's Circulating Library and the Victorian Novel.* Bloomington: Indiana University Press, 1970.

Gunther, Albert E. *A Century of Zoology at the British Museum.* London: Dawsons, 1975.

Gwynn, Stephen Lucius. *The Life of Mary Kingsley.* London: Macmillan, 1933.

Hammond, Dorothy, and Alta Jablow. *The Africa That Never Was: Four Centuries of British Writing About Africa.* New York: Twayne, 1970.

Hanno and Thomas Falconer. *Voyage of Hanno.* London: T. Cadell and Davies, 1797.

Harris, Leslie M. *In the Shadow of Slavery: African Americans in New York City, 1626–1863.* Chicago: University of Chicago Press, 2004.

Henderson, Philip. *Swinburne: Portrait of a Poet.* New York: Macmillan, 1974.

Herodotus. *The Histories.* Oxford: Oxford University Press, 2008.

Hills, William Henry. *The Author.* Vol. 1. Boston: Writer, 1889.

Hobson, Richard. *Charles Waterton, His Home, Habits, and Handiwork: Reminiscences of an Intimate and Most Confiding Personal Association for Nearly Thirty Years.* London: Whittaker, 1867.

Hollingshead, John. *Ragged London in 1861.* London: Smith, Elder, 1861.

Holzer, Harold. *The Speech That Made Abraham Lincoln President.* New York: Simon & Schuster, 2004.

Hombert, Jean-Marie, and Louis Perrois. *Cœur d'Afrique: Gorilles, cannibales et Pygmées dans le Gabon de Paul Du Chaillu.* Paris: CNRS, 2008.

Horan, James David. *Mathew Brady: Historian with a Camera.* New York: Crown, 1952.

Humboldt, Alexander von. *Cosmos: A Sketch of a Physical Description of the Universe.* London: George Bell & Sons, 1901.

———. *Personal Narrative of Travels to the Equinoctial Regions of America.* London: H. G. Bohn, 1852.

Huxley, Leonard. *Life and Letters of Thomas Henry Huxley, by his Son Leonard Huxley.* London: Macmillan, 1900.

Huxley, T. H. *Man's Place in Nature: And Other Anthropological Essays.* London: Macmillan, 1894.

Jackson, Julian R. *What to Observe; or, The Traveller's Remembrancer.* Revised and edited by Norton Shaw. London: Houlston and Wright, 1861.

Jeal, Tim. *Livingstone.* London: Heinemann, 1973.

———. *Stanley: The Impossible Life of Africa's Greatest Explorer.* New Haven, Conn.: Yale University Press, 2008.

Johnson, Steven. *The Ghost Map.* New York: Penguin, 2006.

Jordan, David S. *Leading American Men of Science.* New York: Henry Holt, 1910.

Kennedy, Dane. *The Highly Civilized Man: Richard Burton and the Victorian World*. Cambridge, Mass.: Harvard University Press, 2007.

Kingsley, Mary. *Travels in West Africa*. London: Macmillan, 1897.

Larson, Edward J. *Evolution: The Remarkable History of a Scientific Theory*. New York: Random House, 2006.

Lewis, Cherry, and Simon Knell. *The Making of the Geological Society of London*. London: Geological Society, 2009.

Lightman, Bernard V. *Victorian Popularizers of Science: Designing Nature for New Audiences*. Chicago: University of Chicago Press, 2007.

Livingstone, David. *Missionary Travels and Researches in South Africa*. New York: Harper & Brothers, 1872.

Lovell, Mary, *A Rage to Live: A Biography of Richard and Isabel Burton*. New York: W. W. Norton, 2000.

Lutz, Deborah. *Pleasure Bound: Victorian Sex Rebels and the New Eroticism*. New York: W. W. Norton, 2011.

MacMahon, T. W. *Cause and Contrast: An Essay on the American Crisis*. Richmond: West & Johnson, 1862.

Marcy, Randolph Barnes. *The Prairie Traveler: A Hand-Book for Overland Expeditions*. New York: Harper & Brothers, 1859.

Markham, Sir Clements Robert. *The Fifty Years' Work of the Royal Geographical Society*. London: John Murray, 1881.

Marsh, George Perkins. *Man and Nature; or, Physical Geography as Modified by Human Action*. New York: Charles Scribner, 1864.

Mathews, William. *Hours with Men and Books*. London: S. C. Griggs, 1877.

McCarthy, Justin. *Portraits of the Sixties*. New York: Ayer, 1903.

McClellan, George. *McClellan's Own Story*. New York: C. L. Webster, 1887.

Meredith, Roy. *Mr. Lincoln's Camera Man, Mathew B. Brady*. New York: Dover, 1974.

Meyer, Lysle E. *The Farther Frontier: Six Case Studies of Americans and Africa, 1848–1936.* Cranbury, N.J.: Associated University Presses, 1992.

Mill, Hugh Robert. *The Record of the Royal Geographical Society, 1830–1930.* London: Royal Geographical Society, 1930.

More, Henry, and René Descartes. *A Collection of Several Philosophical Writings of Dr. Henry More.* London: J. Downing, 1712.

Morris, Ramona, and Desmond Morris. *Men and Apes.* New York: McGraw-Hill, 1966.

New York City Landmarks Preservation Commission. *Guide to New York City Landmarks.* Hoboken, N.J.: John Wiley & Sons, 2009.

Owen, Rev. Richard. *The Life of Richard Owen.* 2 vols. London: John Murray, 1894.

Pardon, George Frederick. *The Popular Guide to London and Its Suburbs.* London: Routledge, Warne and Routledge, 1862.

Paston, George. *At John Murray's: Records of a Literary Circle, 1843–1892.* London: John Murray, 1932.

Patterson, Karl David. *The Northern Gabon Coast to 1875.* Oxford: Clarendon Press, 1975.

Perham, Margery, and J. Simmons, eds. *African Discovery: An Anthology of Exploration.* Evanston, Ill.: Northwestern University Press, 1963.

Perrois, Louis. *The Art of Equatorial Guinea: The Fang Tribes.* New York: Rizzoli, 1990.

Picard, Liza. *Victorian London: The Tale of a City, 1840–1870.* New York: St. Martin's Griffin, 2005.

Pike, Godfrey Holden, and James Champlin Fernald. *Charles Haddon Spurgeon: Preacher, Author, Philanthropist.* New York: Funk & Wagnalls, 1892.

Porter, Charlotte M. *The Eagle's Nest: Natural History and American Ideas, 1812–1842.* University: University of Alabama Press, 1986.

Purchas, Samuel. *Hakluytus Posthumus; or, Purchas His Pilgrimes.* Vol. 6. Glasgow: James MacLehose and Sons, 1905.

Quammen, David. *The Reluctant Mr. Darwin: An Intimate Portrait of Charles Darwin and the Making of His Theory of Evolution.* New York: Atlas Books/Norton, 2006.

Ravenstein, E. G., ed. *The Strange Adventures of Andrew Battell of Leigh, in Angola and the Adjoining Regions.* London: Hakluyt Society, 1901.

Reade, W. Winwood. *The African Sketch-Book.* London: Elder, 1873.

———. *Savage Africa.* London: Smith, Elder, 1864.

Redmond, Christopher. *Welcome to America, Mr. Sherlock Holmes.* Toronto: Simon & Pierre, 1987.

Rhodes, Richard. *John James Audubon: The Making of an American.* New York: Random House, 2004.

Rice, Edward. *Captain Sir Richard Francis Burton: A Biography.* Cambridge, Mass.: Da Capo Press, 2001.

Rich, Jeremy. *A Workman Is Worthy of His Meat: Food and Colonialism in the Gabon Estuary.* Omaha: University of Nebraska Press, 2007.

Rotberg, Robert I., ed. *Africa and Its Explorers: Motives, Methods, and Impact.* Cambridge, Mass.: Harvard University Press, 1970.

Rupke, Nicolaas. *Richard Owen: Biology Without Darwin.* Chicago: University of Chicago Press, 2009.

Schaller, George. *Year of the Gorilla.* Chicago: University of Chicago Press, 1964.

Schoch, Richard W. *Victorian Theatrical Burlesques.* London: Ashgate, 2003.

Secord, James A. *Victorian Sensation: The Extraordinary Publication, Reception, and Secret Authorship of "Vestiges of the Natural History of Creation."* Chicago: University of Chicago Press, 2000.

Sigel, Lisa Z. *Governing Pleasures: Pornography and Social Change in England, 1815–1914*. New Brunswick, N.J.: Rutgers University Press, 2002.

Smith, John Thomas. *The Streets of London, Anecdotes of Their More Celebrated Residents*. London: Richard Bentley, 1861.

Souder, William. *Under a Wild Sky: John James Audubon and the Making of "The Birds of America."* New York: North Point Press, 2004.

Spann, Edward. *Gotham at War: New York City, 1860–1865*. Lanham, Md.: Rowman & Littlefield, 2002.

Spurgeon, C. H. *C. H. Spurgeon's Autobiography, 1856–1878*. London: Passmore and Alabaster, 1899.

Stafford, Robert. *Scientist of Empire: Sir Roderick Murchison*. Cambridge, U.K.: Cambridge University Press, 1989.

Stanley, Henry Morton. *The Autobiography of Sir Henry Morton Stanley*. Boston: Houghton Mifflin, 1911.

———. *How I Found Livingstone: Travels, Adventures, and Discoveries in Central Africa*. New York: Scribner, Armstrong, 1872.

Swinburne, Algernon Charles. *The Swinburne Letters*. New Haven, Conn.: Yale University Press, 1959.

Thackeray, William Makepeace. *The Letters and Private Papers of William Makepeace Thackeray*. Vol. 2. London: Octagon Books, 1980.

Thompson, Arthur Bailey. *The Visitor's Universal New Pocket Guide to London*. London: Ward and Lock, 1861.

Timbs, John. *Curiosities of London: Exhibiting the Most Rare and Remarkable Objects of Interest in the Metropolis with Nearly Sixty Years' Personal Recollections*. London: Longmans, Green, Reader, and Dyer, 1868.

Trollope, Anthony. *The Warden*. London: Longman, Brown, Green, and Longmans, 1855.

Van Tassel, David Dirck. *"Behind Bayonets": The Civil War in Northern Ohio*. Kent, Ohio: Kent State University Press, 2006.

Vaucaire, Michel. *Gorilla Hunter.* New York: Harper & Brothers, 1930.

Walker, André Raponda. *Notes d'histoire du Gabon.* Libreville: Éditions Raponda Walker, 2002.

Waterton, Charles. *Essays on Natural History.* London: Frederick Warne, 1870.

———. *Letters of Charles Waterton of Walton Hall.* Edited by R. A. Irwin. London: Rockliff, 1955.

———. *Wanderings in South America.* London: J. Mawman, 1825.

Weber, Bill, and Amy Vedder. *In the Kingdom of Gorillas.* New York: Simon & Schuster, 2002.

West, Richard. *Congo.* New York: Holt, Rinehart and Winston, 1972.

Wheat, Marvin T. *The Progress and Intelligence of Americans: Collateral Proof of Slavery.* Louisville, Ky.: Marvin Wheat, 1862.

Wilson, J. Leighton. *Western Africa: Its History, Condition, and Prospects.* London: Sampson Low, 1856.

Wollaston, A. F. R. *Life of Alfred Newton: Professor of Comparative Anatomy, Cambridge University, 1866–1907.* New York: Dutton, 1921.

Woods, Robert. *The Demography of Victorian England and Wales.* Cambridge, U.K.: Cambridge University Press, 2000.

Wright, Thomas. *The Life of Sir Richard Burton.* London: Everett, 1906.

Yanni, Carla. *Nature's Museums: Victorian Science and the Architecture of Display.* New York: Princeton Architectural Press, 2005.

Yerkes, Robert. *Almost Human.* New York: Century, 1925.

Index